I0051007

Volume 1

CONSERVATION AT HOME

CONSERVATION AT HOME
A Practical Handbook

MICHAEL ALLABY

Routledge
Taylor & Francis Group

NEW YORK AND LONDON

First published in 1988 by Unwin Hyman Limited

This edition first published in 2020
by Routledge
52 Vanderbilt Avenue, New York, NY 10017

and by Routledge
2 Park Square, Milton Park, Abingdon, Oxon OX14 4RN

Routledge is an imprint of the Taylor & Francis Group, an informa business

British Library Cataloguing in Publication Data
A catalogue record for this book is available from the British Library

ISBN: 978-0-367-43303-1 (Set)
ISBN: 978-1-00-300237-6 (Set) (ebk)
ISBN: 978-0-367-42227-1 (Volume 1) (hbk)
ISBN: 978-0-367-42228-8 (Volume 1) (pbk)
ISBN: 978-0-367-82288-0 (Volume 1) (ebk)

Publisher's Note
The publisher has gone to great lengths to ensure the quality of this reprint but points out that some imperfections in the original copies may be apparent.

Disclaimer
The publisher has made every effort to trace copyright holders and would welcome correspondence from those they have been unable to trace.

Conservation
at
Home

A Practical Handbook

Michael Allaby

Illustrations by Andy Martin

UNWIN HYMAN
London • Sydney

First published in Great Britain by Unwin Hyman,
an imprint of Unwin Hyman Limited, 1988.

UNWIN HYMAN LIMITED
15-17 Broadwick Street, London W1V 1FP.

Allen & Unwin Australia Pty Ltd
Napier Street, North Sydney, NSW 2060, Australia

Allen & Unwin New Zealand Ltd with the Port Nicholson Press
60 Cambridge Terrace, Wellington, New Zealand

British Library Cataloguing in Publication Data
Allaby, Michael
  Conservation at home: a practical handbook
  1. Dwellings — Energy conservation
  I. Title
  644   TJ163.5.D86

ISBN 0-04-640002-8

Edited and designed by John Button
Printed in Great Britain at the University Press, Cambridge

Contents

Introduction

Healthy homes add up to a healthy world

This is a book of hints and ideas. It is not a book on gardening, DIY or health, though these are mentioned, neither is it a textbook. It does explain some scientific concepts, but only those relating to its main subject, the protection of the environment and the conservation of resources and wildlife.

The condition of the environment affects us all, since we are all dependent for our health upon a healthy environment. But beyond voting for the politician with the greatest 'green appeal', what can you or I actually do to improve the environment? What practical steps can we take to conserve resources, or to protect species from extinction? This book may guide your thoughts and so help you to make sensible decisions. It cannot tell you what to do, but it can help you to distinguish real problems from imaginary ones, and warn you about apparent remedies that may turn out to be irrelevant, or even worse than the environmental ills they claim to cure.

Things are seldom what they seem. Over the last quarter of a century we have learned that the world works as a set of complex interlocking systems. What we do today in our own lives can produce effects thousands of miles away, effects which may not be evident for many years to come. And what appear to be simple, straightforward and obvious solutions can often be completely inappropriate, or even harmful.

If that all sounds very vague and general, let me give you an example of what I mean — a true story.

For about two weeks in January, 1987, western Europe, including Britain, was gripped by a spell of extremely cold weather. About two thousand people died in Britain, over and above the number that would be expected to die in two normal weeks.

Clearly they died because of the cold, in most cases because they could not afford to heat their homes adequately. Had their homes been better built and fully insulated, their heating bills would have been lower. That is an environmental issue. But there is more. What actually caused the cold spell?

In summer the land warms much faster than the sea; in winter it cools much faster. Thus in winter the large land mass of Asia cools faster than that of western Europe, which is insulated by the Atlantic. Cold air covers much of Siberia, forming a large area of high pressure held in position by the warmer, less dense air constantly moving from the west, and riding over the cold air where the two air masses meet.

In Britain, most cold spells in winter occur when 'fingers' of Siberian air extend westwards as troughs of high pressure. In 1987, however, the Siberian air was not quite as cold as it is in most years. The difference in temperature between it and the adjacent western air was smaller than usual, which allowed the whole Siberian air mass to move westward, bringing unseasonably warm weather to Iceland and arctic Siberia, and sub-Arctic conditions to Britain.

Why was the Siberian air a little warmer than usual? Scientists at the Climatic Research Unit of the University of East Anglia believe it may be because of the 'greenhouse effect', a general global warming of the climate caused mainly by the accumulation of carbon dioxide in the atmosphere (the 'greenhouse effect' is explained in detail on page 14). And why is carbon dioxide accumulating? Because it is released when we burn coal, oil, peat or wood, and when we clear forests.

If this is so, the deaths of two thousand people in Britain may be due to the way we burn coal and oil, and the way we influence land use policies that lead to forest clearance, mostly in the tropics. Everything, you see, is linked to everything else. The world is much more complicated than we used to imagine, and more paradoxical. Thus a particularly cold spell may be due to global warming.

This is just one example of an apparently natural, unavoidable event that may partly have been due to the way we — you and I — behave. There are many more.

Can we change the world?

Concern about the natural environment is nothing new. There were what we would call 'environmentalists' in Ancient Greece and Rome, worrying about urban pollution and squalor, and the soil erosion caused by large-scale farming. The first British laws to conserve forests date from the middle ages, and so do the first attempts to control smoke from coal fires.

Our present concerns — or the old concerns in their modern dress — began in the 1960s, and before long a popular environmental movement had grown up, drawing attention to a long list of pollutants being discharged into our air and water. In the summer of 1972 the United Nations held a Conference on the Human Environment in Stockholm, which led to the establishment of the United Nations Environment Programme, based in Nairobi. Most national governments created agencies, departments or ministries for environmental protection. If the world was to be saved, the saviours were to be the politicians and salvation was to take place on a vast international scale. To the simple question 'What can I do?' came the simplest of answers: 'Vote for me!'.

It seemed that there was little left that we individually could do or needed to do. The villains had been located in the remote offices of multinational corporations, not at the firesides and in

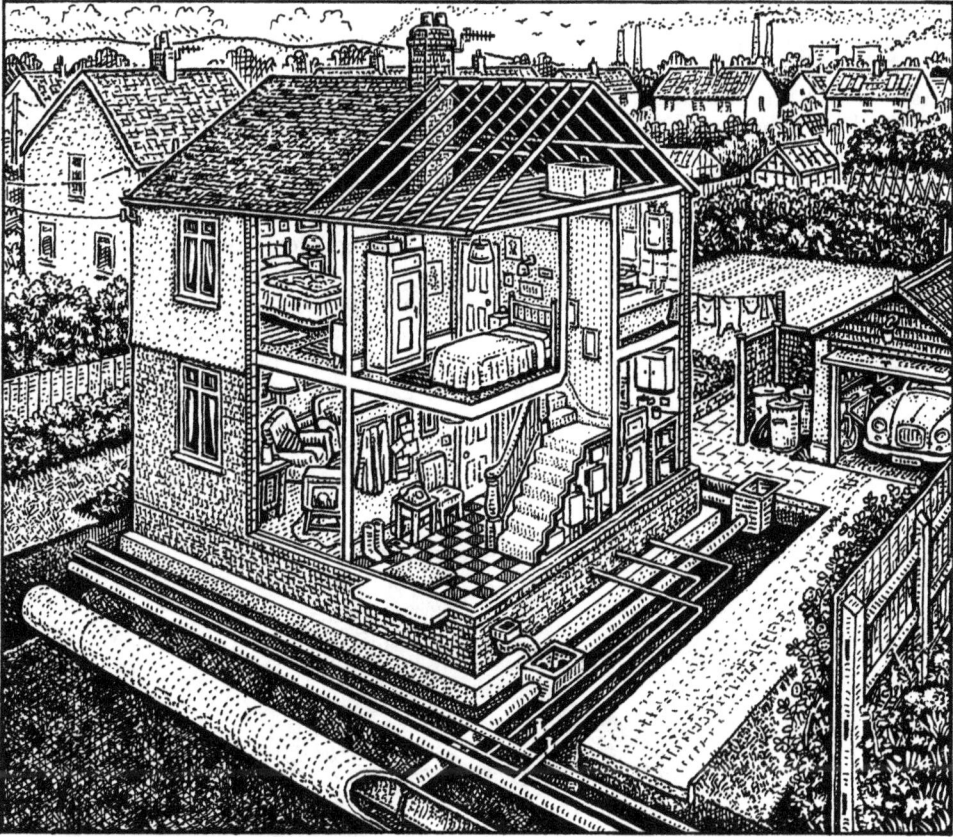

the kitchens of ordinary homes in ordinary streets in ordinary familiar towns. Yet that is exactly where change was really needed. As this book shows, we can start at home, and right now.

Healthy homes can be cheaper

We cannot transform the world overnight, but we are not powerless. We are consumers of products, which gives us economic influence, and we can choose how we use these products. If we spent differently or, even better, spent less, we would inflict less injury on the air, water and wildlife around us. If you are concerned about whether or not a particular product is safe to use, why not start by asking yourself whether you need to use it at all? If you choose not

to use it there is no way that it can harm you or anything else.

This book describes many common household products and the effects of their manufacture and use, and in many cases it questions their real value.

Healthy homes are safer

You and I are part of the natural environment just as much as any other organism that inhabits it, and what hurts others may well hurt us. Since we may spend most of our time inside our houses, this environment is just as important and just as vulnerable as the outside environment. Surely we owe it to ourselves to make it as healthy as possible, and thus to keep ourselves healthy enough to enjoy it.

Conservation begins at home

As you will see, the book consists of many short items, each dealing with a particular subject — a bit like a small encyclopædia. The items are arranged in five larger groups, according to the part of the house or its surroundings in which they are normally found.

Downstairs includes the front door, the entrance hall, the living room with a special section on 'energy', the kitchen with a special section on 'food', the back door, and the pantry.

Upstairs includes the bedroom with a special section on 'fibres', the bathroom, and the lavatory.

The house itself deals with the building, and includes the roof, walls, foundations and drains.

Outside the house deals with the garage and sheds, and with transport.

The garden considers home energy production, food production, weeds and pests, non-edible plants such as lawns and hedges, and wildlife, in the hope that you may come to see your garden as a habitat for fascinating and harmless non-humans.

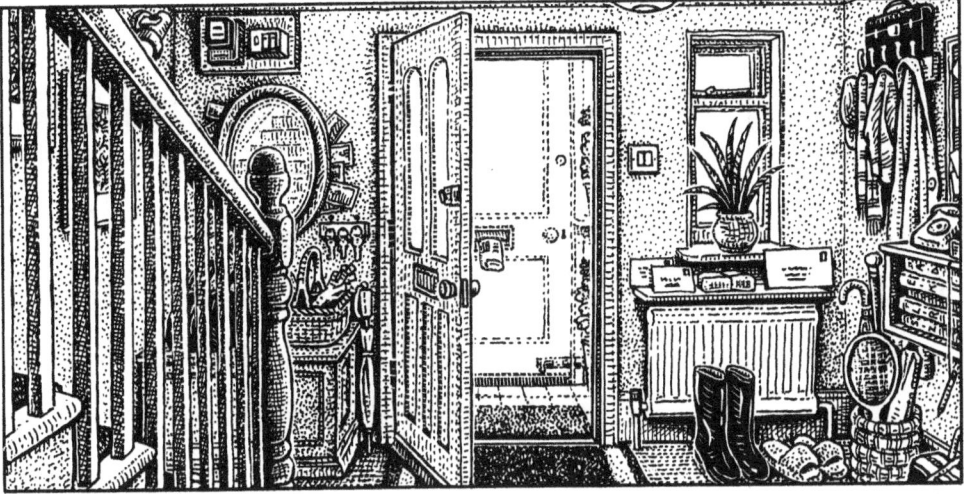

Downstairs

Front door

Draught-proofing Does your front door fit snugly, or does it allow warm air out and cold air in? If it is loose-fitting, draught-proofing strips fitted around it will pay for themselves in a single winter in lower heating bills. Lower heating bills means less work for the power stations, less oil and coal to be mined and burned, and less pollution of the air. While you are at it, you could fit some kind of flap or cover over the letterbox — it too is a source of draughts.

Porch If you fit a small porch outside the house, or a second door inside to form a vestibule, it will act as an air-lock, reducing the exchange of cold air for warm air every time the outside door is opened. Unless it is very elaborate and expensive, a porch will pay for itself in lower heating bills within a few years, and the interior environment will start enjoying the benefits immediately.

Glazing If daylight cannot illuminate the areas just inside the front and back doors, you will need to keep lights on all the time. If there is no window, it is easier and cheaper to put glass in the door itself rather than create a separate window.

Entrance hall

The hall is not really a room, and no one spends much time there, so is there any point heating it? If you fail to heat it, however, the hall will become damp and the rest of the house may be cold.

Heating In most houses the hall opens directly on to the stairwell, a space occupying the central core of the house with rooms on all floors opening off it. People pass through this core but no one spends much time there, so heating it may seem pointless. Yet if it is not heated you will find that in cold weather the marked temperature difference between the heated rooms and the core will force warm air out into the core and cold air into the heated rooms every time a door is opened. A heater in the hall, turned low, will reduce the temperature difference and thus the exchange of air, using less energy than is needed to warm the rooms each time they are chilled. Whenever you save energy, you save money and reduce pollution. If you have central heating, put the main thermostat in the hall and regulate the temperature throughout the house from there.

Preventing damp   If the air in the hall is damp, the air in the stairwell will be damp, and before long the entire core of the house will be damp. This will encourage the fungi responsible for mould on the walls, wet rot and, paradoxically, dry rot (the name refers to the appearance of the affected wood, not to the conditions that cause the problem). Either the house will eventually become uninhabitable and fall down, or you will have to pay for expensive treatment involving the use

of powerful fungicides which you might prefer to avoid. The specialists who treat damage caused by damp are experts at their job, but that is no reason for providing additional work for them. The answer is to keep the air unsaturated so that moisture will not condense on to surfaces, and the way to keep it unsaturated is to keep it warmer than the moist air outside. This can be done by good draught-proofing and a little gentle heating, which in the end will be both cheaper and safer.

Dirt   Feet and moving air constantly carry material from outside the house to the inside. It consists mainly of soil particles, either as dust or carried in water to form muddy marks when the water evaporates. Though it may be unsightly, this sort of dirt is not a health hazard and causes no harm to the environment, so try not to be too obsessive about it. Cleaning, after all, involves more consumption — hot water, detergents, polishes and mops — all of which cost money and use energy.

Living room

The room in which people spend most of their leisure time must be heated for much of the time, and it is while you sit in your winter cosiness that you might take time to reflect on the world outside and your relationship with it. Is your fire polluting the outside air? An electric fire is clean, but what harm is done when electricity is generated? What are the alternatives to coal or uranium for producing electricity, and what environmental effects do they have? And what about the environment within the living room? Are you safe while you relax?

Energy

Heating Before you plan how to heat your home, it is worth reducing to a minimum the heat that escapes from it, though it is important to allow for sufficient ventilation. Prevent draughts by making sure the windows and doors fit snugly. If you are not using an open fire, scrunch up newspapers and stuff

them up the chimney. If you leave it open you are simply providing a hole which carries the air you have carefully and expensively heated straight up into the sky — if you had a hole like that in the ceiling you would soon do something about it! Then think about the sources of heat you may not normally think about — light bulbs, for example, and human bodies. It all counts. Buildings have been designed even for cold climates which need no additional heating of any kind — they use a combination of thorough insulation and carefully-collected sunshine. You are unlikely to reach this standard with an existing house, but it shows what is possible. You may need much less heating than you imagine.

Gas Natural gas, which is mostly methane, burns efficiently, the products of its combustion being water vapour and carbon dioxide. This makes it very clean to use (though see greenhouse effect, page 14). The conventional view is that natural gas and oil are 'fossil' fuels, produced by the partial decomposition of living organisms in

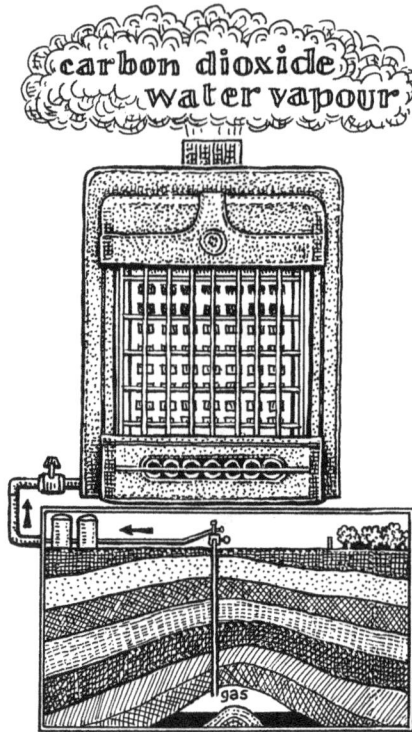

the distant past, and that they are not renewable. One day they will be gone. There is just a possibility that this view may be wrong. Some scientists have suggested that gas and oil are produced deep inside the Earth's crust by natural physical and chemical processes that go on all the time. The gas and oil are sometimes forced upwards, closer to the surface, then trapped in porous rocks beneath a cap of impermeable rock. While they are held in this trap, bacteria start to feed on them; thus rather than being of biological origin, oil and gas are minerals, but with biological contamination. This idea is being tested by drilling for oil and gas in geological structures where biological deposits should not exist, but mineral ones might. If fuel is discovered in such structures it could mean that oil and gas are not 'fossil' fuels at all, and are unlikely ever to be exhausted.

Coal The open coal fire with its cheery flames has two advantages. It is pleasing to look at, and it uses its fuel directly so you benefit from all the chemical energy it contains. Coal can be converted into oil and gas, but this uses energy in the process; if the coal is burned to generate electricity there are further losses, amounting to almost 70 per cent of the original chemical energy of the coal. This might make the coal fire seem efficient, but in fact as much as 90 per cent of the heat vanishes up the chimney. A coal fire is therefore less efficient in use than an electric fire.

Coal is also dirty. It was the burning of coal that made London famous for its filthy buildings, pea-souper fogs, and high incidence of respiratory disease. Over the years these smogs killed untold thousands of people. In the winter of 1952 alone, at least 2000 were killed by smoke-induced smog, and the lives of many more were shortened or made miserable by bronchitis, lung cancer and similar ailments. Our air is cleaner now, but coal-burning is still a health hazard.

Burned coal leaves ash behind, and into the air it emits water vapour,

carbon dioxide, sulphur dioxide, oxides of nitrogen, ash, and radioactive potassium-40 and carbon-14 in amounts up to twelve times those permitted in emissions from nuclear installations of similar power output. Removing sulphur from the exhaust gases leaves behind an acid sludge. If we could give up mining and burning coal, both human health and the environment would benefit.

Finally, coal mining is one of the most dangerous of occupations. All over the world hundreds of coal miners are killed in accidents every year. *See also* acid rain (page 13), greenhouse effect (page 14).

Solid fuel stoves Stoves that burn pellets made from coal are cleaner and more efficient than open fires, but not much. The factories producing the pellets cause serious local pollution, the stoves that burn them often emit irritating fumes into the house, they liberate a fine dust that settles everywhere including your lungs, they are difficult to instal and operate, and almost impossible to regulate precisely. If the stove is used for cooking and hot water it will also heat your house day and night, summer and winter, burning fuel whether you need heating or not. Trendy it may be, but you would be much better off with gas or electricity for cooking and water heating, and so would the environment.

Wood-burning stoves Benjamin Franklin designed the prototype of the 'modern' wood-burning stove, whose popularity grew dramatically in the years when Dutch elm disease supplied large amounts of suitable fuel. A stove like this may look beautiful, but environmentally it is a disaster. By burning wood very slowly with very little air, it produces a range of unburned hydrocarbons, many of which are known to cause cancer, together with carbon dioxide and carbon monoxide, all of which are released into the outside air along with radioactive potassium-40. Pushing this cocktail out into the atmosphere may remove it from your sitting room, but it is no way to treat your neighbours. And now that the supply of dead elm is exhausted it is important to remember that wood stoves burn wood, which comes from trees, which come from the forests that conservationists are supposed to be conserving. If you have a wood-burning stove you could try selling it for scrap, or polish it and put a vase of flowers on it, but on no account should you light it unless you have no other form of heating, and even then you should feel thoroughly ashamed of yourself. It may be the most polluting device ever invented.

Electric fires For the person who uses it, electricity is the cleanest of all forms of energy — it releases nothing. It has to be generated, of course, but there are many ways to generate electricity. It is bound to no particular fuel, and it is much easier to restrict pollution in a large generating plant than in many small installations. In Britain, nuclear and coal-fired power stations provide almost all our electricity, with hydro-electric stations contributing a smaller amount. Hydro-electric stations are clean because they burn nothing, but only some rivers provide sufficient power. Nuclear power is also safe and clean (see below); coal-fired power

stations are relatively dirty and dangerous (see page 8).

Nuclear power In a nuclear power station the decay of radioactive fuel releases heat which turns water into steam. The steam drives turbines connected to generators. The reactor, containing the radioactive material, is sealed from the outside world. The process emits nothing except for small and strictly regulated amounts of radioactive materials. Its wastes are stored while they decay, and although their long-term disposal causes public alarm, the quantities involved are small and techniques for their safe disposal well-understood (*see* nuclear fuel cycle, below; nuclear waste, page 17; radiation, page 16). Discharges from the nuclear power industry account for 0.1 per cent of our total exposure to radiation, and there is no evidence that at this level they are injurious to human beings or that they cause any harmful effect to any other species.

No large industry is entirely safe, but the risk to the public from nuclear power generation, even allowing for the Chernobyl accident, is very small, and Chernobyl was the first civil nuclear accident to cause casualties among the public. The Three Mile Island accident caused no injury, and there was virtually no release of radiation from it.

The Chernobyl reactor is of a type

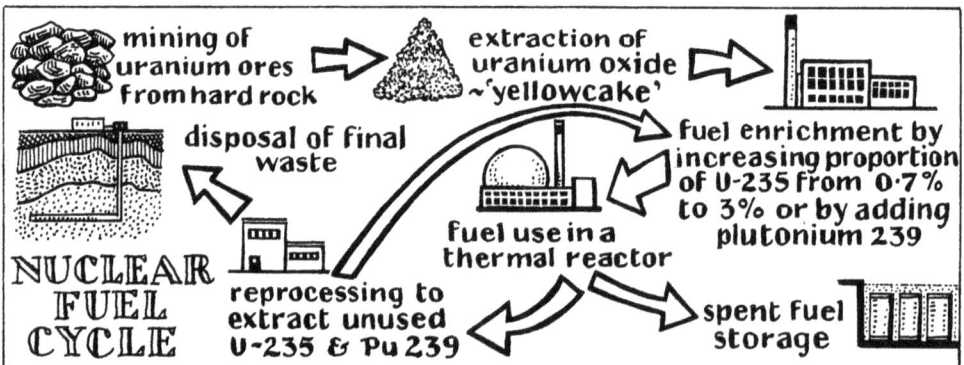

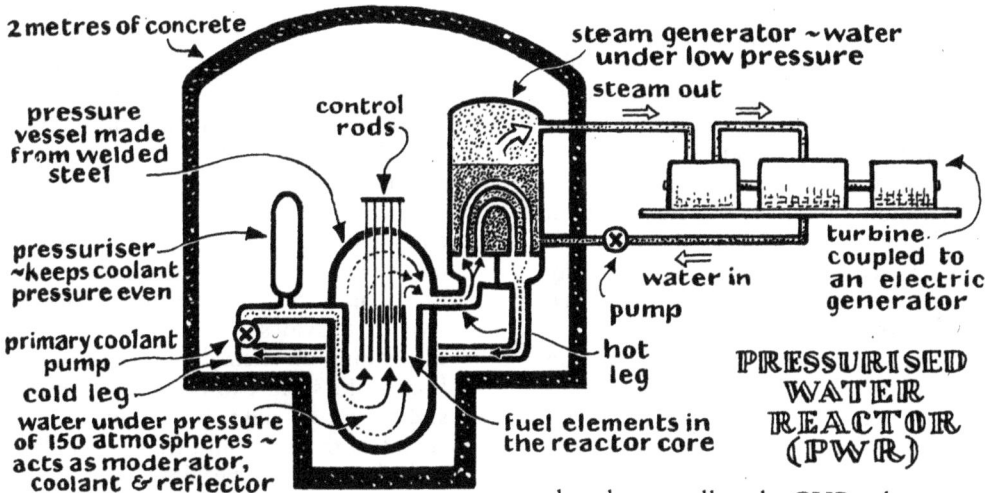

2 metres of concrete

steam generator ~water under low pressure

steam out

pressure vessel made from welded steel

control rods

pressuriser ~keeps coolant pressure even

primary coolant pump

cold leg

water under pressure of 150 atmospheres ~ acts as moderator, coolant & reflector

water in

pump

hot leg

fuel elements in the reactor core

turbine coupled to an electric generator

PRESSURISED WATER REACTOR (PWR)

fundamentally different from any used outside the USSR. It exploded because of a chemical reaction between water and graphite, and no other commercial design uses these two substances in the same reactor. Therefore the Chernobyl accident could not possibly happen in reactors of other types.

The accident killed 31 people directly, injured about 200 more, and may have shortened the lives of several thousands more throughout the northern hemisphere, mainly in the western USSR. Britain was barely affected, at levels that may lead to the deaths of up to fifty people in the next twenty years, but since 40,000 people die from lung cancer alone each year, these additional deaths will be undetectable. The restrictions placed on sheep movements were based on extreme caution, being triggered at radiation levels one-tenth of the international safety limits agreed earlier in 1986.

Combined heat and power Electricity is generated by burning fuel, either to raise steam to drive a turbine or to drive an engine linked to a generator. Burning fuel also produces heat, and CHP aims to use this heat to provide domestic hot water. In a factory, hotel or hospital, or within a small neighbour-

hood, a small-scale CHP scheme can pay for itself in a few years; after that it saves you money. You need a suitable engine and an electronic control system — a car engine adapted to run on gas will do, running at a constant speed and thus dramatically increasing its life-span, though special engines are made for the job. The engine drives a generator connected to the mains electricity supply, but when the engine is running the meter runs backwards! The engine exhaust runs through heat exchangers which feed the heated water to a storage tank. The system is called a 'minichip', and if properly installed it should last about 25 years. The technique is still being developed, and you will probably hear a lot more about it in the next year or two.

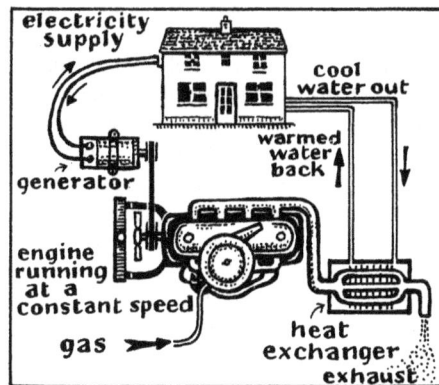

electricity supply

cool water out

warmed water back

generator

engine running at a constant speed

gas

heat exchanger exhaust

Solar power A few years from now you may be able to generate electricity by covering the roof with solar cells, the photovoltaic devices used to power the equipment on satellites. Until now the technique has been expensive and less efficient than conventional power generation, but recent developments in California have reduced the cost and improved the efficiency, so one day they may be suitable for individual buildings. Solar cells harness sunlight; solar panels, which you often see on houses, harness solar heat and use it to heat water. Beware of them! They are expensive to buy and install, and depend upon such complicated plumbing that you should not try making your own unless you are a plumber. *See* solar panels and solar cells in the section on the roof, page 59.

Wind, waves, tide It is tempting to suppose we might use the force of the wind, the swell of ocean waves, or the mighty flow and ebb of the tides to generate electricity from an inexhaustible supply of raw energy. It would be possible, but nothing in this life is free, and in the case of these 'natural' energy sources the cost is rather high. Wind generators need reliably windy sites, which means putting them in open, exposed areas, preferably on high ground — on upland moors, for example, or on coastal clifftops, both of which are usually areas of great scenic beauty. Wind generators are large, standing more than 40m tall, with two or three aircraft-propeller-like blades more than 18m long. Each generator produces 1 to 1.5 megawatts, so between 650 and 1000 wind generators would be needed to match the output of a conventional power station. They would have to be spaced so they did not interfere with each other, and would occupy a very large area. They would be linked by cables with the electricity grid, and most people would find them ugly. They would be noisy,

and the public could not be allowed on the site because of the danger of a blade failing. There are interesting alternative designs with the rotors turning on a vertical axis rather than a horizontal one, but these are no smaller or quieter, or more efficient.

Wave power devices exploit the rise and fall of sea waves, and could be sited off the Outer Hebrides or in the Western Approaches, where the Atlantic swell provides ideal conditions. Depending on their design they might float at the surface or be attached to the seabed out of sight, but in either case arrays of them would extend for miles, and they would be linked to shore installations by cable. The Western Approaches lie at one end of the busiest shipping lanes in the world, which might be a problem, and repairing storm damage to wave devices might prove difficult.

Tidal schemes are based on barrages built across estuaries, such as has been proposed for the Severn, where there is a large tidal range. The incoming tide flows through the dam, in some designs turning turbines as it does so. Then the dam is closed and the water held in an artificial lake until the tide has ebbed, when the gates are opened and the water flows out, turning the turbines. Tidal barrages alter the flow of water in the estuary, affecting the way that mudflats and sandbanks form. This is bound to affect the billions of small animals that bury themselves in the mud and sand, and the wading birds which depend on them for food. The cost of electricity generated by wind, waves or tides is about the same as that of electricity generated by conventional methods.

Heat pumps If you put your hand behind a refrigerator or freezer you will feel warm air. Heat is being taken from inside and released outside. A heat pump uses the same principle. A chemical compound is pumped around

refrigerant gas
~ammonia or
Freon 12

condenser
coils~gas
compressed
so condenses
into a liquid
and releases
latent heat

water or air
passing over
the coils,
circulates
the heat
given off

electric supply

pump &
compressor

expansion
valve

coils set in
the relative
cold ~ in
the ground
or water
from a
stream

evaporator
coils~liquid
gas expands
and turns into
into vapour and absorbs latent heat

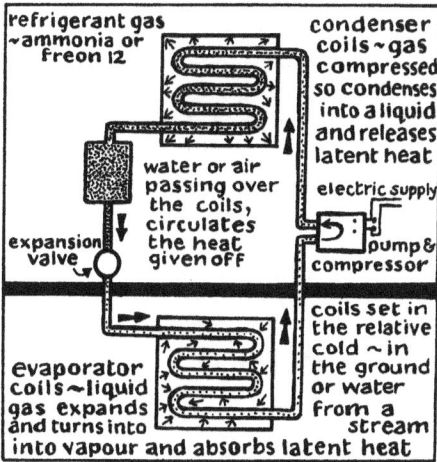

a closed system of pipes. As it enters the area to be heated it is compressed, which makes it condense into a liquid and release latent heat. When it enters the cold area it expands, vaporizes, and absorbs latent heat. In fact it uses a cold area to heat a warm area. Install a heat pump between your living room and the loft, larder, or any other relatively cold area, and it will provide heat just for the cost of running the circulating pump and the compressor. It is said to be the only device whose efficiency is more than 100 per cent. You will find suppliers in your *Yellow Pages*.

Fuel cell For more than a hundred years scientists have been tantalized by what should be the simplest and most efficient of all methods for generating electricity: the fuel cell. Recently there have been further advances in its technology, so the day may come when you could install a fuel cell in your own home. The device is a box with an electrode at each end. Oxygen is passed across one electrode and hydrogen or some other fuel across the other, but the fuel and the oxygen do not come into direct contact with one another The fuel loses an electron, the oxygen gains one, and a flow of electrons — an electric current — is produced between the electrodes. The fuel cell is silent,

produces very little waste heat, and is highly efficient no matter how large or how small it is. Watch out for domestic fuel cells, and when you hear about them take them seriously. They could solve many problems.

Pollution caused by energy production and use

Acid rain The atmosphere contains small amounts of sulphur dioxide and carbon dioxide, emitted mainly by volcanoes, and oxides of nitrogen formed when lightning discharges provide enough energy to oxidize a little atmospheric nitrogen. Water vapour reacts with these gases, and they dissolve slightly in water droplets, forming acids; thus all rain is slightly acid. When we burn fuel containing carbon or sulphur we release carbon dioxide and sulphur dioxide, and when we burn anything at a high temperature, nitrogen oxides are formed. Thus industrial activity increases the amount of acid-forming gases in the air, and the rain becomes more acid.

When groundwater flows over limestone rocks the water becomes slightly alkaline, and can absorb acids with little alteration to its chemistry. Where streams and rivers flow over hard rocks such as granite, however, the water tends to be acid, and the addition of more acid can alter its quality. Many fish are highly sensitive to quite small increases in acidity, and contaminated lakes often lose all their fish.

Vegetation may also be harmed. Rain usually just washes the leaves and has little effect, but fine mist coats the leaves with moisture, and if this moisture is unduly acid it can damage them. Acid particles carried in dry air can also be deposited on leaves and on the ground, where they can affect the chemistry of the soil.

Damage has occurred to lakes in

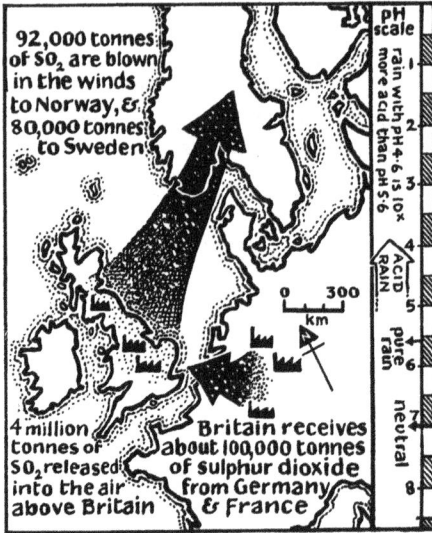

92,000 tonnes of SO₂ are blown in the winds to Norway, & 80,000 tonnes to Sweden

4 million tonnes of SO₂ released into the air above Britain

Britain receives about 100,000 tonnes of sulphur dioxide from Germany & France

pH scale
rain with pH4·6 is 10x more acid than pH 5·6
ACID RAIN
pure rain
neutral

0   300
km

Scandinavia and to a lesser extent in parts of Britain, and to forests, mainly in central Europe. The cause of this acidification is far from clear. Seaweeds emit sulphur compounds into the atmosphere, and for reasons that no one understands (though possibly linked to the greenhouse effect) these emissions have increased substantially in recent years in the North Sea and Baltic. The damage to lakes may be caused partly by industrial emissions of sulphur dioxide and partly by afforestation with conifers, which also makes soils more acid. More than half the industrially-emitted sulphur dioxide comes from coal-fired power stations, though British emissions have been falling recently, having peaked in 1970. We now release about 4 million tonnes a year, which is lower than at any time since the mid-1940s. About 40 per cent of the sulphur dioxide deposited in Southern Norway comes from Britain, but Norwegian soils also contain large amounts of sulphur and there has been rapid conifer afforestation. Air from Britain does not travel to central Europe, and forest damage is unlikely to be caused by sulphur dioxide. The air in German forests contains very little sulphur dioxide, and lichens that are sensitive to very small amounts of sulphur grow abundantly.

The cause of damage in central Europe is more likely to be nitrogen oxides reacting in sunlight with hydrocarbons to release ozone. The trees may also be suffering from disease and the effects of the severe drought of 1976, the year when the damage was first noticed. Nitrogen oxides and hydrocarbons are released mainly by motor vehicles, and their effects are local, thus German forests are probably being most harmed by German drivers. British emissions of nitrogen oxides have remained fairly constant since the early 1970s, at around 1.8 million tonnes a year.

Greenhouse effect Radiation from the Sun, with a wavelength of 400-700nm (1 nanometre = one thousand-millionth of a metre), passes unimpeded through the atmosphere. It warms the Earth's surface, which then radiates heat at wavelengths of 800-4000nm. Carbon dioxide absorbs radiation between 1200 and 1800nm and radiates it in all directions, so warming the air. When the air is warmer, more water evaporates into it. Water vapour also absorbs radiation, especially at around 2000nm, warming the air still more. Methane, chlorofluorocarbons and some other compounds also absorb long-wave radiation, and together, these gases trap heat. Most scientists believe that the climate of the Earth is now warming, and that we may expect major climatic changes by the middle of the next century.

Greenhouse effect Whenever we burn wood, peat, coal, gas or oil, carbon dioxide is released into the air. It is also

1. Radiation from the sun with a wavelength of 400~700 nanometres

Earth's surface warmed radiates heat at wavelengths of 800-4000 nanometres

2. long-wave radiation absorbed and then re-radiated by chlorofluoro-carbons, methane etc. — carbon dioxide — water vapour — so warming the air

3. Increased amounts of carbon dioxide in the air leads to warmer air and so warmer oceans with more water vapour evaporating into the air.

water vapour

carbon dioxide

Burning wood, gas, oil, coal & peat, with deforestation, releases carbon dioxide into the air.

4. sunlight reflected back into space

Increased evaporation causes more clouds to form and more rain to fall

Warmer seas could cause the polar ice sheets to melt and sea levels to rise.

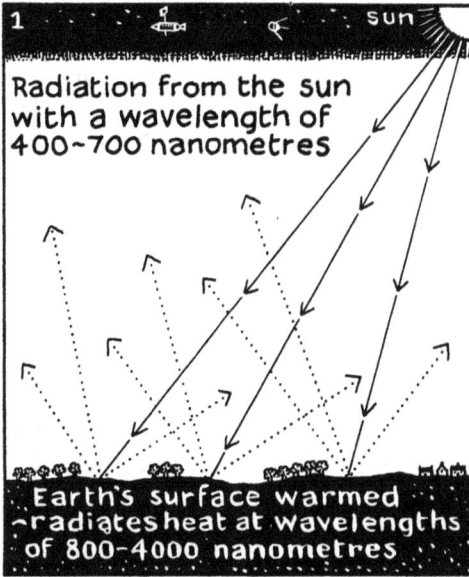

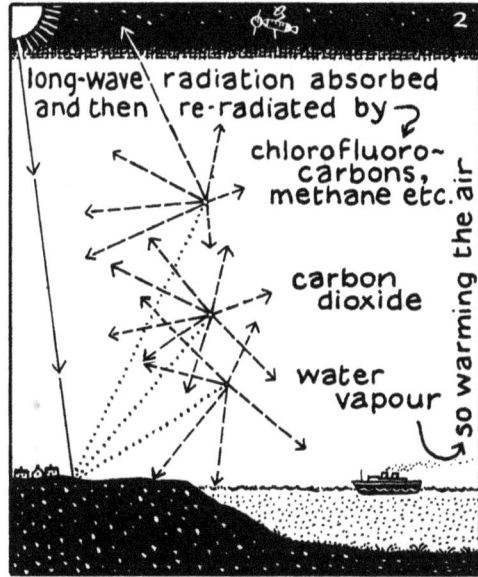

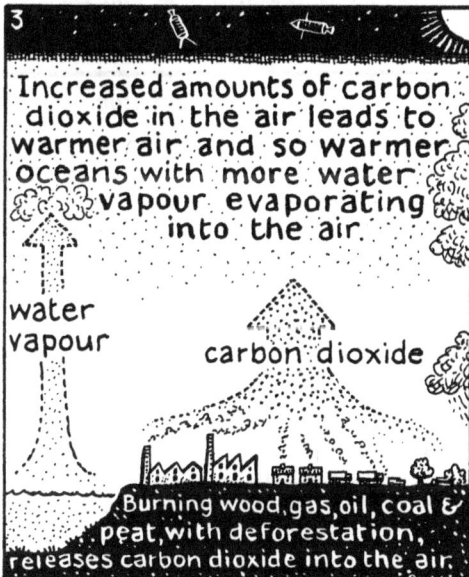

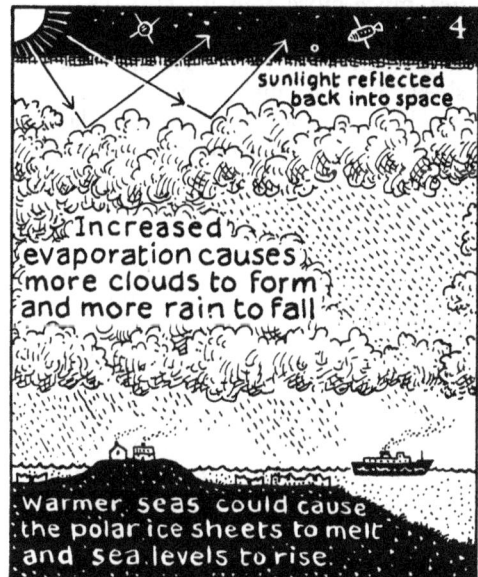

released when forests are cleared, because air trapped among soil particles around the roots of trees contains large amounts of carbon dioxide.

About half the carbon dioxide that enters the atmosphere is taken up by growing plants or dissolves in the oceans. The remainder simply accumulates, though very slowly. In the fifteenth century the air contained about 270ppmv (parts per million by volume) of carbon dioxide; today it contains about 350ppmv. Like some other gases (*see* box), carbon dioxide traps heat in much the same way as a greenhouse.

If the air becomes warmer, the oceans are also warmed. This has two consequences. Increased evaporation puts more water vapour into the air and water vapour traps heat, but more clouds also form. Clouds shelter the Earth's surface and cool it, and the clouds themselves reflect incoming sunlight, which also has a cooling effect.

Nevertheless, on balance the world climate seems to be heating. There is now evidence that both air and oceans have gradually been growing warmer for several years. This warming has been blamed for the droughts in Africa, and it may be responsible for the cool wet summers in Britain in 1985 and 1986, since when the Atlantic is particularly warm this is the result. It may also be responsible for the sharp cold spell in Europe in January 1987 (*see* introduction, pages 1 and 2).

Sea levels are rising because the seas are warmer, but if the warming continues, much more dramatic changes may follow within the next century or two. If the Greenland and Antarctic ice-sheets were to melt, rising sea levels would inundate low-lying land. England would become a small group of islands; Belgium, Holland and much of the eastern USA would disappear.

This is an extreme forecast, and there is little possibility of a runaway greenhouse effect that would render the planet uninhabitable, but most climatologists predict major changes in climate by the middle of the next century.

In the end we may have no alternative but to reduce very substantially the amount of fuel we burn, partly by making economies, but mainly by relying more on nuclear power, whose generation releases no carbon dioxide or other 'greenhouse gases'.

Radiation Radiation occurs either as energy waves (electromagnetic radiation) or as particles. Depending on its frequency, electromagnetic radiation ranges from radio and television waves, through microwaves, heat, infra-red radiation, visible light and ultraviolet light, to X and gamma rays. Only

gamma and X-rays are seriously dangerous. Particles may be helium nuclei (consisting of two protons and two neutrons and called alpha particles), electrons or positrons, called beta radiation, or neutrons.

Alpha and beta radiation from outside the body cannot penetrate ordinary clothing and skin, but can cause damage if emitted from a radioactive substance that has been inhaled or swallowed.

Neutron radiation occurs only inside nuclear reactors or bombs, and is not associated with radioactive wastes or fallout. We are constantly exposed to radiation, from space (cosmic radiation), and from radioactive elements contained in many rocks, seawater, and most organic matter.

Sources of radiation to which we are all exposed are set out in the box.

Sources of radiation	
Natural:	
radon	33%
gamma (from space and from the ground)	16%
internal (from food and our own bodies, as potassium -40 and carbon-14)	16%
cosmic	13%
Total natural	78%
Artificial:	
medical	20.7%
occupational	0.4%
fallout	0.4%
miscellaneous	0.4%
discharges from industry	0.1%
Total artificial	22%

People in the parts of Britain most heavily contaminated by fallout from Chernobyl may have increased their annual exposure by up to one-sixth.

If all this worries you, try growing Tradescantia (Spiderplant) as a houseplant. Radiation makes the cells in its flower stamens change from blue to pink. Examine the stamens under a strong lens or microscope and you can see whether or not you are being irradiated.

Nuclear waste Radioactive waste substances are classified — according to the amount, intensity and type of radiation they emit as 'high level', 'medium level' and 'low level'.

Low level waste consists of gloves, overalls, masks and other items of clothing worn by workers, laboratory glassware, and similar slightly-contaminated material. Much of it comes from hospitals, laboratories and factories, though some is from nuclear power stations. It is mixed in such a way as to be only slightly more radioactive than ordinary soil or rock, and is not dangerous. It is suitable for dumping at sea in deep water beyond the edge of the continental shelf, or burying on land.

Medium level waste comes from nuclear power stations. It is dangerously radioactive but cool, and once stored it requires no further attention. It must be isolated from people until its radioactivity has decayed to a level equal to that found in ordinary uranium-bearing rock. Depending on its content, this may take up to a thousand years. Techniques for its safe storage are well-known.

High level waste is intensely radioactive, and hot. It consists of such materials as spent fuel from nuclear reactors, together with the metal casings in which fuel is held inside a reactor core. It is sealed in steel tanks which are held in ponds below water for about ten years, by which time the waste has cooled. The tanks are then removed, placed inside a thick metal casing, and held in a well-ventilated store where the flow of air cools them further. After fifty to seventy years they are cool and their radioactivity has decayed sufficiently for them to be classified as medium level waste ready for permanent storage.

Using energy efficiently

Double-glazing As a device for keeping out unwanted noise, double-glazing is fairly effective. As an energy-saving device it might reduce your heating bills by one-twentieth. Set the saving against the cost of installing it in an existing house and you will see that it may take something like an entire human lifetime just to break even. Double-glazing only makes sense where a new house is being built, or where the windows of an existing house need to be replaced for any other reason. *See* radon (page 20), ventilation (page 18).

Lighting An incandescent light bulb works by passing an electrical current through a filament, which heats the filament up until it glows. The bulb becomes hot, so much of the energy is wasted as heat. A fluorescent strip light generates no heat, making it cheaper to operate. This suggests that the bulb is inefficient, but all may not be as it seems. When you need the light on the temperature is often low — because it is dark outside, or a very dull day — so the additional warmth may not be wasted after all. If you used strip lighting instead would you need to turn the heating up just a little?

Any saving you make will be trivial, since incandescent bulbs use very little energy anyway. It takes a hundred-watt bulb ten hours to consume one unit of electricity. A better economy might be to invest in long-life bulbs. They are rather expensive, but are guaranteed to last a very long time, and use less electricity than a conventional bulb.

Unless it is wearing out, faulty, or badly-installed, a modern strip light should not flicker. Even so, many people find its very white light harsh and unpleasant.

It is not true, incidentally, that you can cause permanent damage to your eyes by reading in a poor light, even though that is what we were all told as children. The worst that can happen is that reading can be difficult and you will tire yourself. Your eye muscles may hurt, but that is because you have been straining them in your efforts to focus. Neither they nor the eyes themselves will be injured.

The indoor environment

Ventilation In most rooms, badly-fitting doors and windows and an open chimney allow the air to be changed several times every hour. You need to bear this in mind when you seal all the cracks or your room will grow stuffy, though not for the reason most of us imagine.

With every breath we take we remove a little oxygen from the air and very slightly increase the proportion of carbon dioxide. You might think this would be the reason for the increased stuffiness, but in fact human beings cannot affect the chemistry of the air to the extent where we might notice it.

The reason for the feeling of stuffiness is that heat is not being carried away from your body efficiently enough — the discomfort comes from overheating. This in turn will almost certainly be due to a high humidity level which reduces the rate at which perspiration evaporates from your skin. When we breathe we are also exhaling water vapour, and it is this, not the carbon dioxide, that is the problem (*see also* humidity control, below).

Air can also smell stale because odours are not being removed.

You can open a window, of course, but you can also calculate the rate at which the air in a room needs to change. One complete change every hour is usually sufficient, but if the room is very smelly it may need up to six changes an hour. You can work it out more precisely, and may need to do so if you are planning to buy an extractor fan, since fans vary in power. In a room containing not less than 14 cubic metres of air per person (provided there are no very strong smells like cooking smells or someone smoking) air should be changed at the rate of about 0.2 cubic metres per minute (*see also* radon, page 20).

Humidity control   Our homes were once almost always too damp, and keeping them dry was the problem. Coal, gas and oil fires all release water vapour into the air. Now that we use electric fires and central heating, no water vapour is released, and our homes are often too dry for comfort, and may also allow static electrical charges to accumulate (*see* below). We thus need to put some moisture back into the air. You can buy humidifiers to do this for you, but there is a much cheaper alternative that works just as well even if it looks less elegant. Fill a shallow tray with layers of cloth (old blanket or towel is ideal), place it somewhere warm (but not in a cupboard), soak the cloth in water, and

keep it soaked. If this is not enough, fill more trays and find other places for them.

Static electricity   When gamma or X-rays, or fast-moving electrically-charged particles, travel through the air they impart energy to the air molecules they meet. This strips electrons from the molecules, which then have a positive electrical charge. They are 'ionized', and some of their lost electrons flow to the ground. An electric current is simply a flow of electrons. The air is being ionized all the time, most strongly near the top of the atmosphere in the region called the 'ionosphere'. Though the Earth itself is a good conductor of electricity, air is a poor conductor, so an electrical current flows with difficulty from the air to the ground. If there is a better conductor, and almost anything solid in contact with the ground will do, the current will take a short cut. It will flow through a human body, though it is so weak that you cannot feel it and it has no effect. Modern homes sometimes interfere with this current. Water is a good conductor, so if you dry the air you reduce its conductivity. A wall-to-wall carpet is a good insulator, forming a barrier between the electrons and the ground. The air is still being ionized, but the flow of current is reduced; if you also are trapping air in a draught-proof room, so preventing the exchange of charged air with uncharged air from outside, a charge will accumulate. It is 'static' electricity because too few electrons flow to the floor to reduce the charge. If you wear shoes with insulating rubber or composition soles, the current can no longer flow through you, and atoms at the surface of your body may be ionized. When this happens, your charged hair stands on end in its efforts to discharge itself into an uncharged comb passed through it, and you may hear little crackling noises, or even see sparks and electrons from ionized air molecules flow to the Earth ~with difficulty through air, but more easily through a building or a person in contact with the ground

a static charge of electrons accumulates on a body insulated from the ground, in a room with dry still air

rubber soles fitted carpet

static charge cleared by ventilation, moist air, leather soled shoes or no shoes

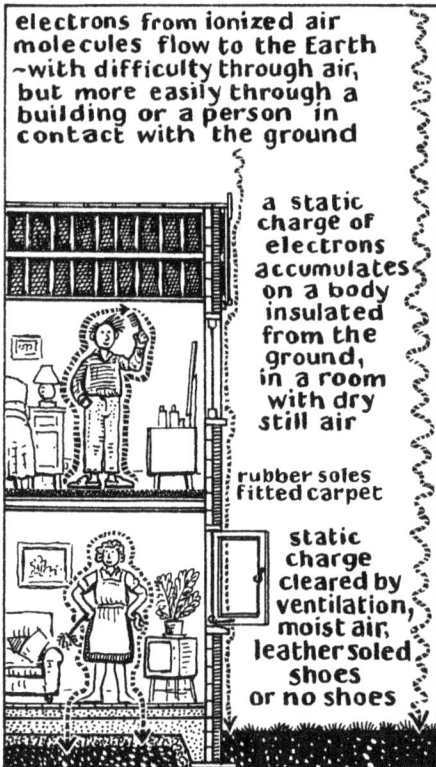

feel small electric shocks. Some people are sensitive to a static charge. It can cause headaches, a vague feeling of unease, and other symptoms. If this is your problem you may be tempted to rush out and buy an air ionizer to produce negatively-charged air molecules in the hope that they will cure you. They may, but there are much cheaper ways of clearing a static charge and preventing it from building up again. If your symptoms are acute, take a shower or bath to clear the charge from your skin and hair. Indoors, wear shoes with leather soles or no shoes at all, rather than rubber or composition soles. Open the window to let in air from outside, and keep the air moist (*see* humidity control, above).

Radon A colourless, tasteless and odourless gas, radon is an element formed and emitted when radium-226 decays. Radon continues to decay,

emitting alpha radiation and forming several other short-lived radioactive elements which tend to stick to solid objects. Small amounts of radium are present in almost all rocks and soils, so radon is present everywhere. It is the largest single source of our exposure to radiation, on average accounting for 33 per cent of our total exposure. The actual amount of radon varies widely from place to place according to variations in local geology, and is generally higher where the underlying rock is granite.

Radon is short-lived, the most stable of its isotopes having a half-life of about 3.8 days. It enters our homes mainly through the floor from the underlying soil, but also from the outside air, and from the walls themselves if these are built from stone or brick. There have been fears that improved home insulation, which reduces the frequency with which inside air is exchanged for outside air, might allow radon to accumulate indoors, but research has not confirmed this. Some homes do have relatively high radon levels, but it makes no difference whether they are draughty or well-insulated.

Radon can cause lung cancer, but the doses to which most of us are exposed are too low to pose a serious threat. It has been estimated that a lifetime's exposure to the average level of radon produces a risk of lung cancer fifty times lower than that faced by the average cigarette smoker. However, because a risk exists when levels are higher than average, the British Government has set limits. If the annual indoor radiation level exceeds 20 millisieverts, buildings must be modified by sealing doors and sometimes walls, and new buildings have to be designed to prevent radiation levels exceeding five millisieverts a year. If you suspect that the radon level in your home may be high, ask the local authority to arrange to have it measured. This service is free, but you will have to pay for any necessary alterations.

Television Some people are afraid that their TV sets may emit harmful radiation. Until about twenty years ago some sets did emit radiation, but the problem was recognized and designs altered to prevent emissions.

A television tube is a glass funnel from which most of the air has been evacuated. At one end there is an 'electron gun'. Controlled by the radio signal received by the aerial, it uses a very high electrical voltage to fire (from a cathode, hence 'cathode ray tube') and direct an intense stream of electrons along the tube to scan a special coating on the back of the screen, making it glow.

Cathode ray tubes are also used in much the same way to produce X-rays, and an old, dirty, badly-maintained colour television tube may emit a tiny amount of X-ray fluorescence. This radiation is not intense enough to cause any harm, and a modern tube that is working properly emits none at all.

Chemicals

Decorating materials If used correctly, modern paints, wallpapers and pastes will not harm you or the environment. You must not swallow them, of course, though even if this should happen paints are safer than they used to be, though most of them contain fungicide (*see* fungicide in paint, below). Arsenic is no longer used as a green pigment, and lead is not used in paints intended for household use. Emulsion paints consist of a pigment mixed with a colloidal suspension of a plastic polymer, all held in water. These paints dry by the evaporation of the water, leaving the pigment bound to the polymer. The paint itself is not poisonous, nor are the fumes released from it. Gloss paints use the same pigments, but they are held in a synthetic resin, usually with other substances added to accelerate drying and ensure the even dispersal of the pigment. The fumes released as the paint dries can be irritating, but they cause no permanent harm. Wallpapers are made either from paper or from polyvinylchloride (PVC), and are harmless (though you should not burn PVC because it then produces poisonous fumes). Paste is a harmless cellulose derivative. When you have finished decorating, seal the paint tins securely, and either keep them or put them in the rubbish for disposal. Do not pour paint down your drain, though the white spirit you wash your brushes in and the unused paste will do no harm; neither will washing your brushes in running water.

Fungicides in paint Most paints and wallpaper pastes contain fungicides to retard the growth of moulds, which can be common on damp walls. There is nothing that you can do to remove the fungicides, though once your decorating is finished they cannot harm you. They are safe while you are decorating as long as you do not swallow them, but you should always wash your hands thoroughly when you finish. The quantities you are likely to release into the outside environment when washing out brushes or buckets are too small to cause any harm.

Polychlorinated biphenyls (PCBs) and mercury When electrical equipment is broken or worn out, be careful how you dispose of it. Never burn it, because it contains plastics that emit poisonous fumes. You can throw it out with ordinary household refuse, but try not to throw very much away at one time — it can cause pollution. PCBs are chemical compounds closely related to the organochlorine insecticides (which include DDT), and are used in electrical insulation, especially in transformers.

Chemically very stable and persistent, they are harmless to humans, but in the outside world they can accumulate, just like organochlorines, until they reach concentrations high enough to injure wildlife. Serious pollution has occurred in the past due to the discharge of industrial waste contaminated with PCBs into lakes and rivers. The tiny batteries used to power modern watches and other small devices may contain mercury. In Japan people have been throwing away these batteries in such numbers that there are fears of serious mercury pollution from rubbish tips.

Insect pests

Cat and dog fleas If your pet is infested you may be able to see the fleas jumping around, or one may even dine on your own blood. In Britain, cat and dog fleas transmit no disease to human beings and are merely an annoyance which disappears of its own accord in late autumn, when most of the fleas die and the rest become dormant. The best way to get rid of them is by vacuuming

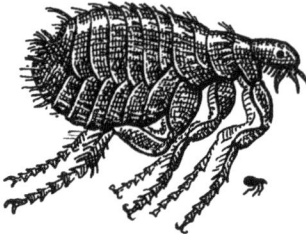

furniture and floors, especially carpets, and washing the bedding where your pet rests or sleeps. Comb an infested cat daily using a very fine-toothed comb — this should remove fleas and their droppings (which are food for the larvae). If the infestation is severe (as it is in some years) you can use an insecticide, though choose one containing methoprene, which poisons fleas without poisoning cats, dogs or humans. Beware of flea collars for cats. They can

catch on projections, injuring the cat; they can cause dermatitis; and the cat may suffer from pesticide poisoning, to which all cats are especially sensitive.

Woodworm In severe cases of woodworm the wood is completely destroyed — if this is the case it should be removed and destroyed at once before other wood is infested. If the damage is mild, insecticide can be squirted into all visible escape holes to impregnate the galleries and kill any late-developing larvae. The most-used insecticide is a mixture of lindane and pentachlorophenol; lindane is fairly harmless to humans, but pentachlorophenol is very poisonous if inhaled or swallowed, irritates the eyes and skin, and can penetrate the skin. Read the instructions on the container carefully, follow them precisely, and take no chances. You can avoid the possibility of woodworm by having exposed but out-of-the-way timbers treated. Other wood should be as smooth as possible to remove the small indentations and cracks in which the eggs are laid, then you can apply gloss paint, varnish or polish to make a surface on which females can find nowhere to lay their eggs.

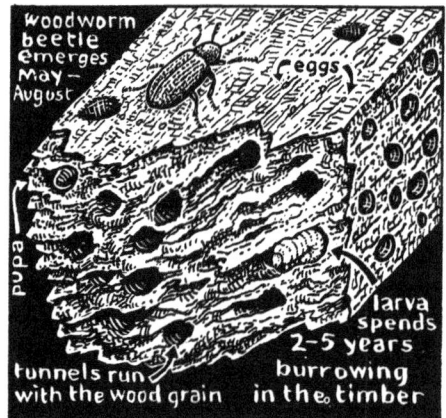

woodworm beetle emerges May–August — eggs — larva spends 2–5 years burrowing in the timber — tunnels run with the wood grain — pupa

Kitchen

The room where food is prepared is the appropriate place to think about the world food situation, and whether or not our eating habits influence it. This leads to the methods by which food is produced in this country, and thence to the ways we prepare and cook our food.

The kitchen is also the place where dishes and clothes are washed. Washing involves machinery, which uses energy and chemicals. Are those chemicals harmful to us, or to the environment that receives them when we pour them down the sink?

What effect is your kitchen having on the global environment?

World hunger

Imported food Is it true that we import vast amounts of food from poor countries, thus depriving them of it and contributing to world hunger? In fact, Britain is close to self-sufficiency in all the foods that can be grown in our climate, and the bulk of our imported food comes from other EEC countries, or from developed countries such as the USA, Israel, Australia and New Zealand.

Our rice comes mainly from Europe or the USA, soya beans from the USA, and maize from Europe. From Third World countries, Britain imports cane sugar (under the Commonwealth Sugar Agreement), some tropical fruits, tea, coffee, cocoa, and small amounts of some other commodities.

The small amount that we do import from the Third World is paid for in sterling or some other international trading currency which the exporting countries need badly to finance their imports of essential materials and technology. If anything, we import too little of any real value from the Third World, though I am not suggesting that we should buy food which they need to feed themselves or products which use land that could be feeding people.

Eating habits and world hunger Would there be more food available to feed the poor of the world if we in the West ate less meat and more cereals, beans, and other protein-rich plant foods? It would probably make very little difference.

We could change our eating habits so we had more to spare, but the farmers who grew it would have to be paid, and the cost would be passed on to the consumer. The food would have to be sold, and the very poorest Third World countries could not afford to buy it. Our government could pay the farmers and then give the food to the hungry. This is an excellent way to relieve emergencies, but when it was tried over a long period the results were disastrous. In the recipient countries the free or nearly-free food was used in preference to locally-grown food, for which farmers had to be paid the full local price. Local farm incomes fell, farmers could not afford to buy tools and seed so their production fell, and far from helping local agriculture the scheme came close to bankrupting it, which would have left people perpetually dependent on food aid — in effect on handouts. They would never have been able to feed themselves.

Hunger is rarely due to a shortage of food. It is caused by poverty. There is — or could be — food available, but people have no money to buy it. If we want to solve the problem of world hunger, we must address ourselves to the alleviation of poverty. It is the world economic and trading systems that we need to change, not our eating habits.

Meat-eating If you prefer a vegetarian diet, and balance it sensibly, there is no reason why you should be short of any nutrients. If you are a vegan, eating no animal produce whatsoever, you may need to supplement your diet with iron and calcium, and you will certainly need to take vitamin B12.

It may be true that the raising of animals for food involves cruelty, and that modern farming methods produce meat with more saturated fatty acids than are good for us. It is not true, however, that meat-eating is in some way unnatural. Physiologically, human beings are omnivores, adapted to digesting meat. Anthropological and archaeological evidence shows clearly that our ancestors ate meat, and that many of them ate a larger proportion of meat than we do. Cereals were domesticated only relatively recently, so if anything it is the consumption of cereals that is unnatural.

On the other hand, it is true that in many parts of the world meat production is a wasteful use of land. The reason is quite simple. If the land is capable of growing plant crops that human beings can eat directly, then human beings will receive the whole of the food grown on that land. If the plants are fed to animals and human beings then eat those animals, this is no longer the case. Like human beings, farm livestock use most of the food they eat to provide them with energy; when we eat animals we do not eat the whole of them, so there is waste. An efficient livestock system can provide human consumers with about ten per cent of the food value of the original plant crops, but many systems are far from efficient. Thus if we grow plants, feed them to animals, then eat the animals, at least nine-tenths of the original food is lost to us. If you feed people on the plants, you can feed many more of them.

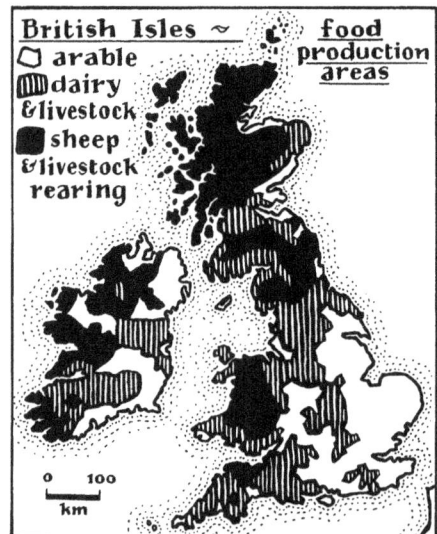

British Isles ~ **food production areas**
arable
dairy & livestock
sheep & livestock rearing

0   100
km

This is a strong argument in favour of vegetarianism, but it does not really apply in Britain, because in Britain this is not the way that most meat is produced. Large areas of Britain are climatically unsuited for growing anything other than grass, and although it is possible to process grass into a food we can eat, the resulting food has no flavour, colour or texture — these have to be added artificially, so the end product is both expensive and not very appetizing. As it is, sheep are raised wholly on grass, beef cattle almost wholly, and dairy cattle mainly. Pigs and poultry are fed on cereals, some of which is fit for human consumption, so we might eat less of them. If we did though, prices of other meat would increase, poor people would suffer, and the grain mountains would grow even larger.

The land-use argument applies in countries where it is no more difficult to grow vegetables and cereals than it is to grow good pasture, and this amounts to vast areas in low latitudes. Obviously we should not encourage people living in those latitudes to produce meat to export to us, though apart from a little corned beef from Argentina, Britain mostly imports meat only from other EEC countries.

Food quality

Food additives Salt, pepper, sugar, spices, herbs and gravy browning are all additives, and most food additives are very necessary. Antioxidants, for instance, are used to prevent oils from oxidizing and becoming rancid — when they go rancid oils can produce poisons, some of which are cancer-forming. Other preservatives are used to prevent food from spoiling between harvest and the time it is eaten, and colours are sometimes added to make food more appetizing and thus more digestible. Most colours are taken from plants, and are therefore quite 'natural'.

These days, however, additives are usually called by their chemical names, which people find unfamiliar. The flavour of apples is caused by acetaldehyde, and of pears by propyl acetate. In Europe, additives are given codes, the 'E' numbers, rather than scientific names. The system was meant to help, but has only made the confusion worse. E300 is simply vitamin C, for example, and E150 is caramel, the colour in gravy browning, but give them an E number on a label and who but a chemist is to know?

The safety of additives used by the food industry is subject to constant and rigorous scrutiny, but there is a problem. The chemical industry can mass-produce additives, so it strongly encourages their use regardless of whether they are necessary or desirable. Apart from the obvious waste, some additives cause illness in people sensitive to them, so we should urge the food industry to use additives only where it can be shown that food is improved by them, and that the consumer really benefits.

Some 'health foods' are of doubtful quality. If they really contain no preservatives, they may be contaminated and even unsafe. Some 'health food' peanut butter, for example, has been found to contain aflatoxins, a group of poisons released by a fungus, at levels considered unacceptably high, because the manufacturers did not do all they could to reduce it. Some aflatoxins have been shown to cause cancer in laboratory animals.

We cannot abandon all additives, and our health would certainly suffer if we tried to. The important thing is to retain a sense of proportion. Do not be persuaded by the chemical industry that every product they need to dispose of is good for you but do not listen either to the scientific illiterates who think that banning additives would make food more 'natural'. 'Natural' and 'unnatural' have no meaning in relation to food; a food that was really 'unnatural' would be inedible. 'Junk food' is also a meaningless expression. Either something is food or it is not.

Nitrates, nitrites and nitrosamines in food Nitrate from excessive fertilizer use is sometimes present in certain foods, especially spinach. Saltpetre (potassium nitrate) or Chile saltpetre (sodium nitrate) are used as preservatives, especially for 'cured' meats such as ham and bacon. They are not added to proprietary baby foods (*see* clean water, page 38). Nitrate used in curing is reduced to nitrite, and nitrite is often used directly. It improves the colour and flavour of the meat and at the same time kills bacteria, most importantly *Clostridium botulinum*, which produces one of the most potent of all poisons and causes botulism. The nitrite also combines with amines which are naturally present in the meat to form nitrosamines. Very large doses of nitrosamines cause cancer in experimental animals; thus the use of nitrates and nitrites is strictly regulated even though they have not been proved to cause cancer in human beings. The incidence of stomach cancer, which nitrosamines might be expected to cause, has in fact been falling for some years.

Hormone and antibiotic growth promoters To make farm animals grow faster and produce the lean meat that customers prefer, some farmers have for some time been feeding chemical growth stimulants to their animals. These stimulants consisted first of small doses of antibiotics added to the feed; later on hormones were implanted in the ears of the livestock, with artificial ones gradually replacing natural ones. The use of antiobiotics used to treat infections in humans might have led to the development of strains of resistant bacteria, thus making diseases associated with them more difficult to treat.

The use of antibiotics is strictly regulated to avoid such health risks, but the regulations have led to a black market in illegal drugs. Hormone growth promoters were investigated very thoroughly by independent scientists who found they presented no risk whatever to human health. They are implanted in the ear, which is not eaten, and traces of them are easy to detect in the unlikely event of meat being contaminated. Unhappily the EEC banned their use without even studying the scientific report it had commissioned, so there may now be a risk that farmers will illegally continue to implant hormones, probably in muscles where they are more difficult to see but may survive long enough to be eaten by the human consumers.

Pesticide residues Some pesticides are poisonous to humans. When farmers and growers use them, they are not allowed to harvest the crop until sufficient time has elapsed for the pesticides to disappear. If, however, they use more than the correct amount of pesticide, or harvest too soon, residues can reach the consumer, and this risk is taken seriously. Produce is sampled and checked for residues both during growth, in the shops, and after cooking, and the regulations and testing techniques are under fairly constant review. Residues are sometimes found, but they are at the lower (and very sensitive) limit of detection, and cannot possibly pose any risk to human health. There is no record of anyone in Britain having been poisoned by pesticide residues in their food.

Organically-grown food Food produced without the aid of artificial fertilizers is said to have been grown 'organically'. Organic farming and horticulture cause less pollution than their 'conventional' counterparts, because poisons are not used and highly-soluble fertilizers cannot contaminate rivers or groundwater. Organic methods are environmentally sound.

Yields tend to be somewhat lower, although the differences are sometimes exaggerated. Now that Britain produces too much food rather than too little this is less important anyway than it was a few years ago.

The use of the terms 'organically grown' (and 'free range' in the case of eggs) have been defined legally, so it is an offence to misuse them in order to deceive the public. You can trust food bearing those labels to be what it claims.

Cooking

Microwave cookers Heat is electromagnetic radiation with a wavelength longer than that of visible light. When the wavelength is increased still further, to between one millimetre and three centimetres, the waves become short radio waves, or microwaves. This is not ionizing radiation, and nothing emitting or exposed to it is radioactive, but microwaves can penetrate some materials, including foods, and when they collide with water molecules they impart energy to them. This heats the water and thus the food, and because the chances of a collision increase as the waves travel further through the material the warming takes place inside the food.

A microwave cooker delivers an intense beam of microwaves to heat food from the inside outwards. Used correctly, it cooks food thoroughly yet very quickly, and uses very little energy in doing so; it is highly efficient. It will not scorch food to produce a crisp dark outside to meat for example, but this may be nutritionally advantageous, since scorching produces harmful substances.

It may be unsafe to use clingfilm in a microwave oven (*see* food packaging).

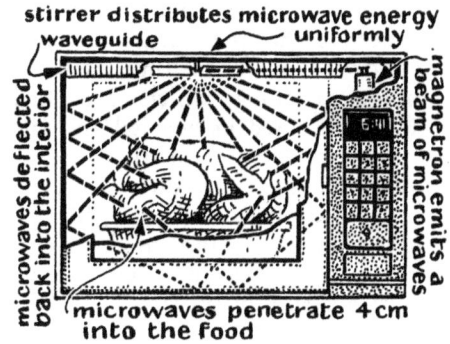

stirrer distributes microwave energy uniformly
waveguide
magnetron emits a beam of microwaves
microwaves deflected back into the interior
microwaves penetrate 4 cm into the food

Microwaves that can cook meat can also injure humans, of course, so the cooker is designed to turn itself off automatically when the door is opened. While it may leak microwaves, these are not very intense or at the wavelength most likely to cause harm. Even so, microwave cookers are often installed just above head height to minimize exposure.

Pressure cookers Probably the most economical of all cooking utensils in terms of nutrient conservation as well as cost, the pressure cooker steams its contents at a pressure rather higher than the outside atmospheric pressure and a temperature higher than the boiling point of water. Food is cooked quickly because of the higher temperature, and few nutrients are lost because little water is used. Because it contains

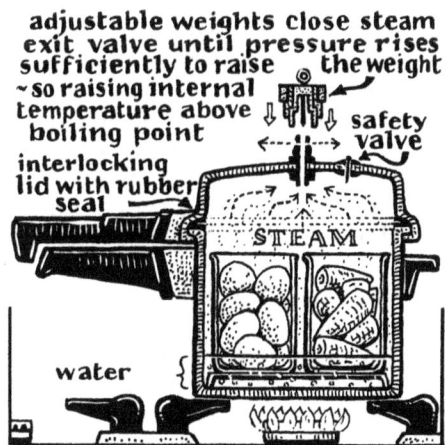

adjustable weights close steam exit valve until pressure rises sufficiently to raise the weight ~so raising internal temperature above boiling point
safety valve
interlocking lid with rubber seal
STEAM
water

its internal pressure, once that pressure has been reached the heat source can be turned very low for the time it takes to complete cooking, which saves fuel. Autoclaves, used in hospital and laboratories to sterilize instruments, are nothing more than pressure cookers, so pressure cooking kills all bacteria present in food, though it cannot remove any bacterial or fungal poisons which are already present.

Pots and pans Can pots and pans contaminate food cooked in them? Some can, but this does not necessarily make them harmful.

Cheap or soft enamel may contain toxic metals, and release small amounts into food if the enamel becomes chipped. Chipped enamel may also provide crevices in which bacteria may grow, and will also produce 'hot spots' on which food may catch. Good, hard enamelware is expensive, but safer. Enamel manufacture involves the use of toxic substances, which can cause pollution around the factory, and a great deal of energy, since the enamel glaze must be fired on to the metal.

Aluminium pots and pans are safe to use. As soon as it is exposed to air, aluminium oxidizes, and the layer of oxide coating it is completely insoluble in water and resistant to most foods. Acids from some fruit may remove it temporarily, allowing aluminium compounds to form in the food, especially — according to some scientists — in areas where the water contains fluoride. The main dietary source of aluminium is tea, since the tea plant concentrates it: Russian tea contains most, then China tea, with Indian tea containing the least. Very high blood levels of aluminium can cause bone and brain damage, and may also be associated with Alzheimer's disease (premature senility), but it is doubtful whether aluminium utensils could contribute enough to cause illness, and Alzheimer's disease is more likely to be

due to the way the body deals with aluminium than with the amount in the diet. Aluminium itself is the most abundant metal on Earth, but its refining uses very large amounts of energy and can cause serious pollution.

Iron pots and pans are very heavy and release small amounts of iron into food, but this simply adds a little to the iron already in our diet and does more good than harm. Iron smelting and steel making are traditional heavy industries and in the past have caused severe pollution.

Stainless steel is resistant to water and to all foods. Toxic metals are used in its manufacture, but to release them into food you would need to raise the temperature to several thousand degrees, high enough to melt the steel. However, stainless steel is a relatively poor conductor of heat, so pots and pans made from it usually have a layer of a better conductor, commonly aluminium or copper, on their bases, but on the outside where the coating cannot come into contact with the food.

Ovenproof earthenware and glassware are also perfectly safe, and are made by a traditional technology which is not a serious source of pollution. Glass does, however, scratch fairly easily, so food can be trapped in small crevices.

Washing and waste disposal

Dishwashers Hardly anyone enjoys washing up, but there it is, a fact of life. Most of us just get on with it. If you are immensely rich, a dishwasher will do some of the washing up for you. It will not wash everything, it will be expensive to buy, and it will use truly impressive amounts of very hot water for washing and hot air for drying — but the dishes will be clean when they leave it.

A dishwasher devotes just as much effort and fuel to a plate with a few breadcrumbs on it as it does to congealed egg and fat, because it cannot tell the difference. A human washer-up is much more efficient. A dishwasher may be convenient if yours is a large family given to eating frequent complicated meals followed by kitchen sink mutinies, but otherwise it is simply a device for disposing of surplus cash.

Washing machines Washing machines are very convenient. If you can afford to buy a machine, life will henceforth be easier, but there is a price to pay. Washing machines use a large amount of energy, most of it in heating water. Even if hot water is supplied by your heating system, you still have to pay for it. The machine uses a lot of water, which is why the water authorities need to build larger reservoirs, install more pipes, pass more water through their treatment plants, and charge you for doing so. Automatic machines use much more water than twin-tubs. The water leaving the washing machine is contaminated with ordinary dirt, along with detergent and fabric softeners.

The best way to reduce this pollution, while also saving energy and water and making your clothes last longer, thereby reducing your bills, is

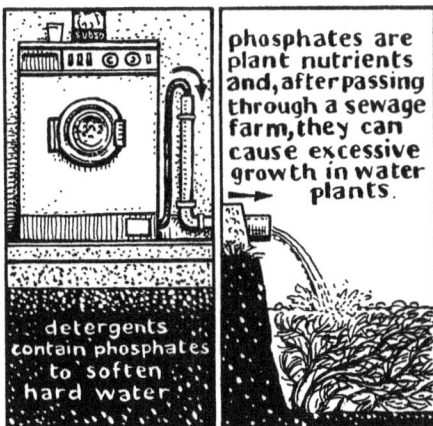

phosphates are plant nutrients and, after passing through a sewage farm, they can cause excessive growth in water plants.

detergents contain phosphates to soften hard water.

to make less use of the machine. Perhaps you do not need a machine at all — hand-washing still works. There is no need to wash clothes unless they are dirty, and you can remove some marks by rubbing or brushing. Choose cooler wash programmes as much as possible; run the machine only when it has a full load. Running an 'economy' load is more expensive per pound of washing than the 'full' load. If you cannot wait to accumulate a full load, trying buying more clothes.

Tumble driers A tumble drier is an extremely efficient device for devouring money, with dry washing as a somewhat incidental by-product. It is bound to be so because it uses energy, and a lot of it, performing a task that can be done just as well, and sometimes much better, using energy that is either free or the by-product of some other activity.

The best way to dry clothes is to hang them in the open air. They will be moved by the wind, dried gently, and the ultraviolet light in sunlight will kill bacteria, so partly sterilizing them. It will also bleach the clothes slightly, so they will look whiter.

If it is raining, which it usually is, you can hang the washing indoors. The interior of a house is one of the driest places on Earth, and a plentiful source of heat. You heat it every time you boil a kettle, and most of the heat generated by human bodies and household equipment warms the air, which then rises. An old-fashioned 'horse' which you can raise to the ceiling on pulleys holds a surprizingly large amount of washing, and dries efficiently because the space near the ceiling is the warmest part of any room. It burns nothing and is much cheaper to buy and install than a tumble drier. A folding clothes horse is even cheaper and simpler, and in winter there will probably be a fire to stand it by, though you must keep an eye on it to make sure it does not catch fire.

Sink waste-disposers When you put scraps of food into a waste disposer there is a loud noise as the monster smashes them to pulp. They are then washed away into the sewer, out of your sight and mind. From there, the slurry goes to the sewage works (if there is one) or straight into the sea (if there is not) to deliver its load of nitrogen and phosphorus compounds. These chemicals can pollute fresh water and are expensive to remove. The ideal place for food scraps is inside chickens which convert them into eggs, or pigs, or a compost heap. The second-best choice is to put them in a loosely-sealed bag in the dustbin. A sink waste disposer is a desperate and expensive last resort.

Chemicals in the kitchen

Aerosol cans An aerosol is a minutely small particle of a substance, so light that it floats in the air for a long time before settling on a surface. If you can break a substance into particles this small, all of them roughly the same size, you can deliver that substance on to a surface very evenly, using very little of it.

To produce an aerosol, the substance must be sealed in a can under high pressure, with a button to release the pressure and so allow the substance to escape through a small nozzle. The pressure is achieved by mixing the substance with a propellant that is liquid under pressure, but expands and evaporates when the pressure is relieved. Various substances will do this — butane (a liquified petroleum gas) is suitable but so inflammable that it can convert a hair spray into a flame-thrower; ammonia works but has a bad smell and is very poisonous; carbon dioxide also works but leaves the can at a temperature so low that it will freeze the surface at which it is directed.

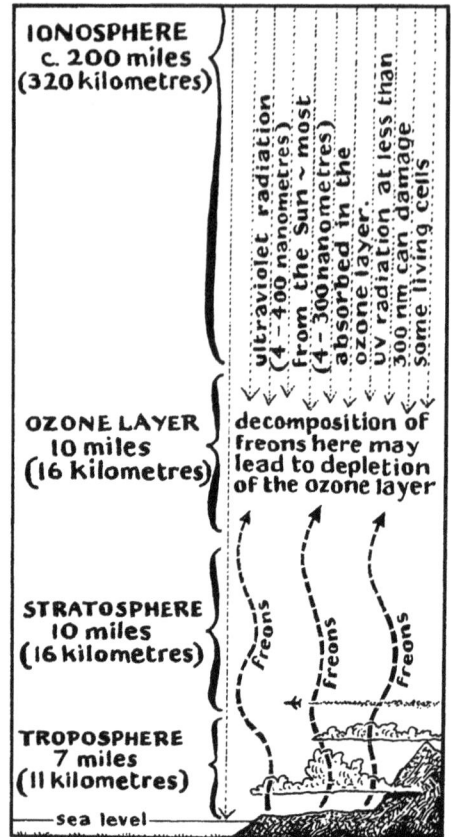

IONOSPHERE c. 200 miles (320 kilometres)

ultraviolet radiation (4 – 400 nanometres) from the Sun — most (4 – 300 nanometres) absorbed in the ozone layer. UV radiation at less than 300 nm can damage some living cells

OZONE LAYER 10 miles (16 kilometres) decomposition of freons here may lead to depletion of the ozone layer

STRATOSPHERE 10 miles (16 kilometres) freons

TROPOSPHERE 7 miles (11 kilometres)

sea level

The most satisfactory propellants are the 'freons' — compounds of chlorine, fluorine and carbon. It is very difficult to make them react with any other substance, so they are non-inflammable (related compounds are used in fire extinguishers) and completely non-toxic. There has been some suggestion that they are serious environmental pollutants, however, because they tend to accumulate in the upper atmosphere, where they may decompose to release chlorine which can form stable compounds with ozone, and so deplete the ozone layer which shields the Earth's surface from much of the ultraviolet radiation emitted by the Sun. This fear has led to the banning of these propellants in some countries, but so far there is little hard evidence of any real risk, and the hazards of exposure to ultraviolet light may have been exaggerated.

Ammonia A useful solvent, especially for cleaning metals, ammonia has an unmistakable smell and is extremely poisonous. Inhaling it can make you cough; inhale enough and it can cause sudden death. Swallow it and it may make you vomit blood; splash it on your skin or in your eyes and it can cause serious injury. It must therefore be treated with great respect, though it is not an environmental pollutant — dissolved in water it is readily converted into an essential nutrient for plants.

Asbestos If you inhale asbestos fibres they may injure you. As with all pollutants, the risk depends on the amount you inhale and the length of time over which you continue to do so. Asbestos fibres cause no harm until they are inhaled, so if you have asbestos around your home (as roofing or lagging for example), it is best to leave it alone. If it is indoors you can cover it with boarding, or give it a generous coat of paint, to seal it so there will be no loose fibres to become detached. Asbestos roofing will release no fibres unless you break or saw it.

You can throw away and replace small items, such as a pad from an old ironing board (new ones do not use asbestos), but if larger amounts are to be removed you should consult an expert (look under 'asbestos removal' in *Yellow Pages*). It is illegal for asbestos to be removed except by a licensed operator.

Biological detergents Many of the substances that stain cloth are bound to the cloth by proteins. Biological detergents contain enzymes (proteases, obtained from a common soil bacterium cultured by fermentation) which break down large protein molecules into smaller fragments, making them easier to remove.

You should not handle biological detergents more than necessary, and some people are more sensitive than others. Human beings are, after all, made largely from proteins, so it is hardly surprizing that some of us are vulnerable to attack.

Biological detergents are reasonably effective, but do you need them? If the clothes have been washed properly they will be clean anyway.

Bleach The most common liquid domestic bleach is sodium hypochlorite, which smells of chlorine, although hydrogen peroxide and sodium chlorite are used to a lesser extent, mainly in industry. Dry bleaching powder, often added to scouring powders, also smells of chlorine. It is a mixture of calcium hypochlorite and calcium chloride. The strength of a bleach is sometimes given as a percentage of available chlorine, chlorine being the bleaching agent.

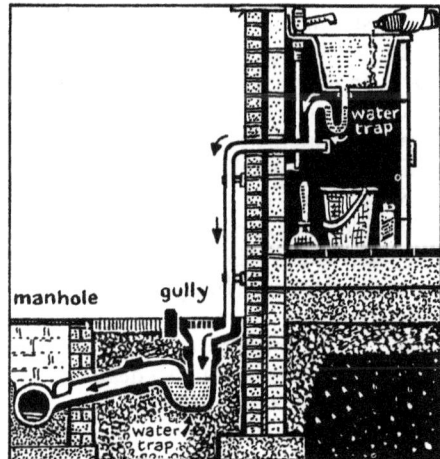

The amount of chlorine given off by domestic bleach is far too small to harm anyone, despite the fact that in large amounts chlorine is very poisonous. Chlorine destroys bacteria, but is inactivated in the process. Far from contaminating the water going down your sink, therefore, it will tend to purify it and then disappear: it is not an environmental pollutant.

Dry cleaning fluids Grease stains can be removed from garments with petrol, or with other solvents used in dry cleaning machines and sold for domestic use. Some years ago the most popular substance was carbon tetrachloride, which is totally non-inflammable, but is very poisonous if swallowed or inhaled. It has been replaced by alternative compounds, but these are also poisonous, though less so, and some are inflammable.

Do not allow a naked flame near any of them, and use them only in a well-ventilated space. If the smell of cleaning fluid gives you a headache or makes you feel nauseous or drowsy, get into the fresh air at once. Because they evaporate so readily, cleaning fluids disperse rapidly in the environment, and are not serious pollutants.

Insecticides in the kitchen Insecticides should not be sprayed or dusted in the kitchen or the larder. If they miss food and utensils you will be wasting your time and money; if they do go on food and utensils you run the risk of poisoning yourself and your family.

Strips hung up to kill insects release pesticide at a constant rate until they are exhausted. They are not likely to contaminate food, but you might prefer not to inhale them. In any case, the insects are almost certainly resistant to most of the products designed to poison them.

Old-fashioned sticky flypapers are safe enough, provided you do not mind the accumulation of corpses, but they kill only those flying insects that alight on them by chance. Insects enter the kitchen in search of food and places to lay eggs. Deny them these by storing food in sealed containers and removing scraps quickly rather than leaving them exposed, and they will leave you alone (*see also* ants, page 35; houseflies and bluebottles, page 36; mice and rats, page 36; silverfish and firebrats, page 37; spiders, page 37).

Methylated spirits Take pure alcohol, add a dye to make it look distinctive, then add some methanol or methyl alcohol (the process is called methylation), and you have methylated spirits. Alcohols have many uses, especially as solvents, which makes them extremely useful for cleaning.

The methanol, which makes methylated spirits very poisonous indeed, is added mainly to stop people drinking it. Methylated spirits disappears very rapidly in the environment, dissolves in water, and evaporates if you spill it. If you pour it down the sink it will cause no pollution, and apart from being poisonous if taken internally and highly flammable, it is safe to use.

Petrol Petrol, including lighter fuel, is a useful solvent for many cleaning jobs about the house. It is highly volatile (meaning that it evaporates quickly) and the vapour is highly inflammable, so use it with caution.

It is poisonous if you drink it, so keep it away from children. Unlike heavier oils, it is not a serious environmental pollutant because it evaporates and disperses so readily.

Soap, detergent, washing up liquid A detergent is a substance that dissolves, usually (though not always) in water, and this property helps to lift dirt from articles being washed. Soap is a detergent. Synthetic detergents, which are chemically very complex, were introduced in the 1940s, since unlike soap they do not react with the calcium in hard water to form a tough insoluble scum.

Early detergents contained alkyl benzene sulphate, which caused spectacular foaming at sewage works since it was not broken down by bacterial action. The formula was changed in the 1960s, and all domestic detergents in use today are 'biodegradable'; they still contain phosphates to soften hard water, however, and these can cause pollution. Phosphates are plant nutrients and stimulate the excessive growth of aquatic plants if they accumulate as they may do in still or slowly-moving water.

Some detergents contain a bleaching agent (often sodium perborate, a dry substance that releases oxygen), but 'whiter than white' effects are achieved by adding an organic chemical that absorbs ultraviolet radiation from sunlight and re-emits it as white light.

You cannot avoid using detergents, but use no more than you need, which is a good deal less than most of us use. It is often possible to wash dishes without using any detergent at all, for example — try it. Detergents vary in strength according to the job they are designed to do. Washing up liquid and shampoos, which come in contact with the skin, are the mildest. *See also* biological detergents, page 31; dishwashers, page 28; washing machines, page 29.

Food storage

Food packaging Food is often wrapped in plastic or sealed in plastic boxes. While the food remains cold it is probably quite safe, but at higher temperatures there is a risk that the substances added to rigid plastics to make them soft — 'plasticizers' — may contaminate food.

The safest food containers are those made from glass, crockery or stainless steel, but the most widely-used wrapping is low-density polythene, which is flexible. Rigid boxes and bottles are made from PVC, high-density polythene or polystyrene, which are satisfactory, but cannot be heated easily and may be slightly permeable. This makes them difficult to sterilize, and allows liquids and gases to pass through them to some extent. Liquids intended for

drinking should not be kept in plastic bottles for more than a couple of weeks. The contents may partially evaporate or become contaminated.

Polystyrene is often used to make the inner lining of refrigerators. Expanded polystyrene is made by blowing an inert gas (usually one of the chlorofluorocarbon compounds or 'freons', among the least toxic substances known) through the compound as it sets.

PVC (polyvinyl chloride), a rigid plastic, is made by treating vinyl chloride monomer (VCM) so the individual molecules join to form chains (the process is called 'polymerization'), but traces of the original VCM may remain. VCM can cause cancer in people exposed over a long period to relatively large amounts, so the amount permitted to remain in

PVC intended for wrapping food is strictly regulated. These days PVC food packaging presents no health hazard.

Clingfilm is PVC to which a plasticizer (di-2-ethylhexyladipate, or DEHA) has been added. If the clingfilm is heated, for example by cooking food inside the wrapping, some DEHA may be released and enter the food. This could be harmful, so you should not cook food in a clingfilm wrapping or use it to wrap hot food.

Food preservation The best way to preserve food is by eating it. Meanwhile keep it cool and covered, to protect it from dust and flies. If it goes bad before you manage to eat all of it, buy less next time.

Milk is traditionally preserved by making it into cheese and fruit by making it into jam. You cannot freeze cows' milk, but you can freeze goats' milk.

Preservation aims to make life impossible for the micro-organisms that cause decay and food poisoning, usually by denying them warmth or humidity or both. Pickling alters the acidity to a level which micro-organisms cannot tolerate. Sugar or salt produce a solution that drains water from cells by osmosis, and so kills by dehydration.

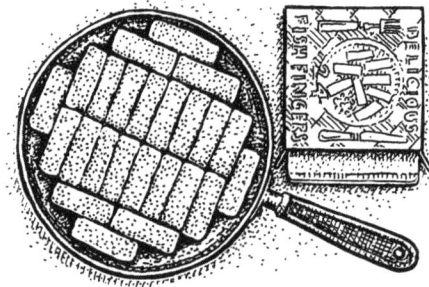

Every form of processing and preservation causes some loss of nutrients, though in freezing and dehydration this is small. Sulphur dioxide is sometimes added to dried fruit, which helps to preserve vitamin C. Canning and bottling, involving heating food inside the container, destroys some of the thiamin, folic acid and vitamin C.

In Britain you cannot yet buy food preserved by irradiation. The evidence so far is that irradiation causes no more nutrient loss than any other form of preservation, can be applied to foods that cannot be preserved in any other way, and is perfectly safe. It does not lead to radioactive contamination of the food.

Freezers If you live in a remote rural area and shop for food three months at a time, then a freezer may be useful. It is also useful if you raise a great deal of your own produce and need to store it. Otherwise it is expensive to run and offers few economies to compensate.

At one time it really was cheaper to buy frozen food in bulk, especially meat, but once freezers became popular, food prices rose to take advantage. Today there is little to be gained; in any case, fresh food, including meat, is always better than preserved food, no matter how efficient the method of preserving.

Refrigerators A refrigerator is nothing more than a food cupboard, although it is efficient because it is airtight. It is not really necessary to keep food chilled, though it does extend the 'shelf life' of some perishable foods by a day or two.

It works by pumping a fluid around a closed system of piping between the inside and outside of the refrigerator, compressing it while it is outside. The fluid condenses into a liquid when compressed, giving up heat, and evaporates when it is allowed to expand, taking heat from its surroundings. The fluid itself is a freon, implicated by some people in damage to the atmospheric ozone layer (there are leakages during the manufacture of the chemicals and of refrigerators), and there is some pressure to ban all freons. If this happens, refrigerators will probably use ammonia, which is satisfactory but poisonous.

If you have no refrigerator, there are other ways to keep food cool. A stone slab in a place away from direct sunlight will be fairly cold. Stand bottles of milk in cold water in a cool place and they will remain cool. Place other foods in sealed containers, cover the containers with wet cloths and keep them wet. As the cloths dry, the latent heat of evaporation will be taken from the container and from the surrounding air.

Kitchen pests

It is obviously best if food is not contaminated by pests, but this does not mean that every invader should be seen as an enemy. The careless use of insecticides can do much more harm than some invertebrate animals. Check first to see if the animal you see is a pest at all, and if it is, can you keep it out of your house without using poisons?

Ants Among the most highly developed of insects, all ants live in colonies with a clearly-defined social structure. They care for their young, some species store seeds for food, some enslave other ants and make them do all the work, some make compost on which they cultivate the fungi on which they feed, still others tend aphids from which they obtain honeydew. In the tropics, colonies of migratory carnivorous army ants will devour every small animal in their path. If they invade a human home the family has to move out until they leave, but can then return confident that the house harbours no other animal, not even a flea.

When British ants come indoors they are usually looking for food. If the scouts find some, the colony will send as large a foraging party as is needed to carry it away to their stores. If they find none they will go away and search elsewhere. If they take to living in the house they will choose somewhere dark and sheltered and you will not see much of them. You may think them a nuisance, but they do little harm even though their food stores and cemeteries are inclined to be a bit unhygienic by human standards.

Do not leave food uncovered, seal small holes and cracks, for example in wainscots, and ants should not trouble you. You can try poisoning them with insecticide if you feel you must. The products sold for this purpose are fairly innocuous, but be careful not to treat food you intend to eat, water containing fish you would like to live a little longer, household pets (cats are especially sensitive), the children, or yourself.

Houseflies and bluebottles Plant and animal wastes, if you leave them alone, will soon disappear as they are recycled by a vast and diverse population of small animals, fungi and bacteria. Houseflies, bluebottles, greenbottles and their many relatives form part of this population.

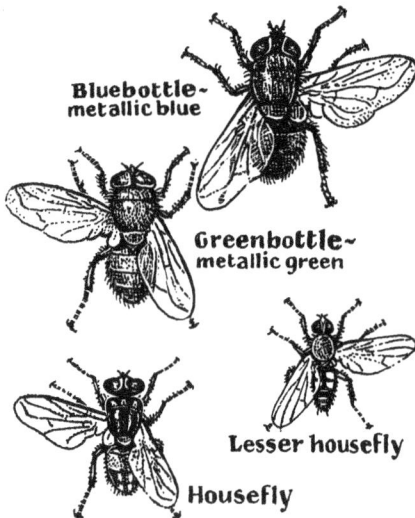

Bluebottle~
metallic blue

Greenbottle~
metallic green

Lesser housefly

Housefly

Two species of housefly are attracted to our homes, along with a few other species that may enter occasionally. They are attracted to the food we leave out for them, which is similar in every important respect to the plant waste they normally live on. When they settle on food to feed on it the flies may leave behind bacteria that they have collected elsewhere, which can cause disease. For this reason people have been trying to kill houseflies for centuries, with very little success. The only way you can discourage them is to keep food covered, and clean up scraps that are spilled. Do not try poisoning them because they are resistant to most insecticides.

Male bluebottles feed on nectar from flowers, and if you see a bluebottle indoors it will almost certainly be a female looking for meat in which to lay eggs. She is not interested in food other than meat, so keep meat out of her way and she is quite harmless. If you find a greenbottle indoors, it is probably there by mistake. It also needs meat on which to lay eggs, but prefers to look for it outdoors.

Mice and rats If rats invade your house you can tell the council about it, or move out, or both, although the brown rat is an intelligent, social, fascinating animal which spends much of its time grooming its fur to keep itself very clean. The dirt we associate with rats was put there by humans. Rats just get the blame for finding uses for our garbage.

There are two British species: the black or ship rat, introduced from ships, probably in the eleventh century and associated with plague, and the brown rat, introduced in the eighteenth century. It spread slowly during the nineteenth century, but is now the more common species.

The familiar grey house mouse was also introduced by humans, but before the Roman occupation. In towns it lives in close association with humans and eats anything they eat. Country mice are more independent and less likely to enter houses.

Mice may transmit diseases to humans and they contaminate much more stored food than they eat. These days they are resistant to warfarin, but be very careful about attempts to poison them. The poison that will kill a mouse may also harm household pets and small children. The presence of a cat will help persuade the mice to leave. Keep food in secure containers, seal small holes and cracks in walls and floors, and if the infestation is serious and persistent seek help from the council. If you find attractive brown mice in the house they will be wood mice or, in some parts of the country, yellow-necked mice. They are likely to leave of their own accord without causing serious problems, though they may be partial to stored apples. *See also* mice, page 36; voles, page 113; and rats in the section on wildlife, page 112.

Mould If you take a sample of air at any time from anywhere on Earth and examine it under a microscope, you will find spores of fungi. When these settle on a suitable medium, they may begin to grow. Eventually they may reproduce, and that is when you will see their furry fruiting bodies. This is mould. So is the *Penicillium* fungus responsible for blue cheese, the yeast *Saccharomyces cerevisiae* used in baking and brewing, and the mixture of species used to make soy sauce and many pickles. Moulds are not all bad!

Fungi are more likely to attack plant products, including cloth, than meats or hides, but there are exceptions. They require ample moisture and a food source to get them started. You can discourage moulds by cleaning or washing clothes before storing them, and making sure that food is clean and really dry before storing it. On no account use fungicides on food.

Silverfish and firebrats Silverfish and firebrats are primitive ancient animals, sometimes called bristletails because their 'tails' are covered with tiny bristles. They are wingless insects covered with silver scales and have tapering bodies up to two centimetres long, with long antennae and three 'tails'. The species have been around for so many millions of years it seems

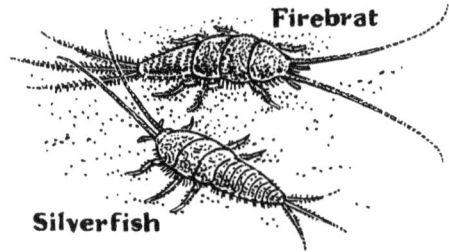

Firebrat

Silverfish

philistine to do anything but admire them. Firebrats live in warm places and their antennae and 'tails' are longer than those of silverfish, which live in the corners of cupboards. Both species feed on spilled flour, glue, bits of paper and similar unconsidered trifles. They are quite harmless, but if you keep those inaccessible dark corners nice and clean there will be nothing for them to eat.

Spiders Most spiders make webs, though not all of them are of the beautiful orb design. Indoors the webs are unsightly and collect dust, so most people remove them.

Spiders themselves, however, are wholly beneficial, and all British spiders are completely harmless to humans. Fierce carnivores, they feed on insects, including houseflies, earwigs, cockroaches and silverfish. They do not attack humans or contaminate food, so when you see one in the bath or scuttling across the carpet, either remove it or leave it alone, but do not harm it. It is a useful ally.

The most common house spider and builder of cobwebs is *Tegenaria gigantea*. The female is relatively large and hairy, the male smaller but with longer legs. He is the spider you sometimes find in the bath. He has not climbed up through the plughole, but has fallen into the bath accidentally while searching for a mate, and is trapped by the smooth steep sides. He will drown if he is immersed in water.

Water

Hardness and softness of water Rainwater is slightly acid. If it flows over hard rocks such as granite on its way to the reservoir, it will remain acid. This is 'soft' water. If it flows across some other rock types, then carbon dioxide dissolved in the water may react with substances in the rock to produce mineral salts, making the water 'hard'.

'Hard' water may contain calcium hydrogen-carbonate and magnesium carbonate as well as iron and other metallic salts. Carbonates are insoluble and form a scale on pipes and in utensils, especially when the water is heated. Because the salts are easily removed from the water, this is called 'temporary hardness'. Metallic sulphates and chloride are not deposited in this way, so they cause 'permanent hardness'. There is evidence from many parts of the world that heart and circulatory diseases of all kinds are less common in hard water areas than where the water is soft, and that where naturally hard water is artificially softened, heart and circulatory diseases increase. No one knows why this should be true, but the link seems clear.

Lead pipes In old houses, and the older areas of some British cities, water pipes are made from lead. This is the main source of the lead we consume. Soft water is slightly acid and reacts with the lead, while hard water deposits a layer of mineral salts over the lead and so prevents direct contact between the water and the metal. Contamination therefore occurs mainly in soft-water areas. If your water is soft and you have lead pipes, do not use water that has stood for a long time in the pipes for drinking or for cooking. If you have not used the tap for a few hours (first thing in the morning for instance), run the cold tap for a few minutes before you take water to use. That will empty the pipes and draw in fresh water that will not have time to absorb any lead.

Water filters Some small water filters are marketed with the claim that they will remove all impurities from your water. In fact they remove nothing at all and are not even effective as water softeners. These small filters will not remove bacterial or chemical contaminants from spring or well water.

Water purity British tap water is clean, wholesome, and perfectly safe to drink. Apart from bottled water, water from other sources (wells included) will be more or less contaminated by industrial or chemical wastes (including fertilizers and pesticides) and bacteria, and its quality will change from season to season. You should not choose it in preference to a mains supply.

All water naturally contains some dissolved nitrate, but the amount can increase when heavy rain washes fertilizer from fields and into rivers. High nitrate levels have been recorded in parts of south-east England and East Anglia, in areas devoted to intensive arable farming, or where treated sewage is discharged into rivers from which water is extracted downstream. Water treatment plants do not remove nitrates, but warnings should be issued when levels rise above the limit of fifty parts per million.

Nitrate is a plant fertilizer, of course, so if you water the garden with nitrate-rich water the plants will benefit.

Babies less than about six months old can be harmed by nitrates, which are converted into nitrites in their stomachs. Nitrites can then pass into their bloodstream, combine with blood haemoglobin, and so interfere with the transport of oxygen by the blood. If high nitrate levels are recorded, it is best for nursing mothers to use only bottled water for themselves and their babies. Older children and adults are not harmed by occasional exposure to high nitrate levels, although prolonged exposure may lead to stomach cancer. This has not been confirmed, but nitrate remains under suspicion.

Chlorine is added to water during purification to kill bacteria, and excess chlorine is removed by dechlorination before the water enters the public supply. Save for the occasional times when dechlorination is inadequate, tap-water contains almost no chlorine, and any it does contain evaporates as soon as the water mixes with air, as it does when it is poured.

In some areas water contains up to one part per million of fluorides, and in a few other areas sodium fluoride is added to bring the content up to this level. High concentrations of fluorides cause staining and mottling of tooth enamel, but at low levels fluoride is harmless, and probably helps reduce tooth decay.

Brown discoloration of tapwater is caused by iron, and is harmless. It occurs when sediment, accumulated over years in mains pipes, is disturbed and flushed through the system. This can happen, for example, when the supply is turned off and then restored.

Water softeners   Most of the salts that make water hard can be removed by adding calcium hydroxide (lime) or sodium carbonate (soda ash, washing soda, or bath salts).

Domestic water softeners use one of two common processes. Some filter the water through a bed of 'zeolite', a hydrated aluminosilicate from whose crystals the water has been driven by heating. This leaves an open crystal structure that captures the salts. This is the technique used, for example, by Permutit. The alternative is to filter the water through polysulphates, such as Calgon, which trap calcium. Eventually the filter medium becomes saturated and must be replaced or regenerated. Domestic water softeners are effective,

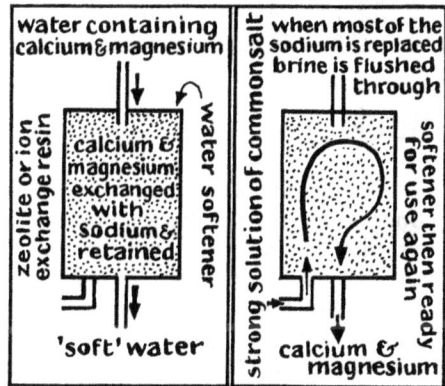

and remove substances from water rather than adding anything, so they are quite harmless. While soft water is most useful for washing, hard water is probably heathier for drinking (*see* hardness and softness of water, page 38).

Back door

In this section we are not so much concerned with the back door itself, but with the area just outside it, where you put the rubbish collected from around the house. What happens to your rubbish when you leave it outside to be collected? How harmless can you make it?

Bottles In Britain we recycle about six per cent of our bottles. Some countries recycle thirty per cent or more. If your town has a bottle bank, use it. If it doesn't, ask the council to open one.

Most recycled glass is used as aggregate in road building, which reduces the need for quarries and gravel pits. If you feel energetic, you could use some bottles to help insulate a greenhouse (*see* greenhouse heating and ventilation, page 80).

Dustbins The rubbish you throw in the dustbin is likely to end its days on a dump made in a hole in the ground. The rubbish gradually fills the hole, so the technique is called 'landfill'.

Or it may be incinerated. In Coventry, Edmonton (London), Jersey, Mansfield, Newport (Isle of Wight),

Nottingham and Sheffield, the rubbish is used as fuel. It may be made into pellets, or simply dried and burned as in an incinerator, but the heat is used to warm buildings or generate electricity. If your council does not use its rubbish sensibly, ask why not. Tell your councillor to get in touch with the Warmer Campaign for Warmth and Energy from Rubbish, 83 Mount Ephraim, Tunbridge Wells, Kent TN4 8BS.

Food scraps Now we put all our rubbish into plastic sacks, food scraps can cause problems at the rubbish dump. When sealed in the airtight bags and buried the food rots, producing methane. If the methane ignites the dump will explode — it has happened, in Derbyshire. People living nearby are not amused, saying it reduces the value of their houses and demanding compensation. Compost food scraps, or cook them into a mash to feed to the chickens, or — if you have to — put them in the rubbish in a loosely-closed bag. *See also* sink waste disposers, page 30; composting, page 83; chickens, page 96; pigs, page 98.

Incinerators When you burn organic — carbon-based — material, the main combustion products are water vapour, carbon dioxide, sulphur dioxide, soot and ash. If the material burns slowly, however, a wide range of other products is released. Blue smoke consists of unburned compounds of carbon and hydrogen, many of which are suspected of causing cancer. Do not burn refuse if there is any other way to dispose of it.

On no account burn plastics. They can release even more dangerous products, including (if they contain chlorine, as PVC does) carbonyl chloride, which smells of freshly-cut hay but was known in the first world war as phosgene, and dioxins. Dioxins are blamed for all kinds of mysterious illnesses. In high concentrations they can cause violent skin rashes (chloracne) as they did when released at Seveso some years ago, but there is no clear evidence that they have any long-term effects.

If you must burn refuse, use an incinerator that allows a good flow of air to the fire rather than a bonfire heaped on the ground. Material in an incinerator will burn faster, hotter and more completely, which means more cleanly.

Plastic refuse There is little you can do with empty plastic bottles except throw them in the dustbin (though *see* alternatives to chemicals, page 88). From there most of them go to a dump, where they can cause problems. They contain pockets of air and usually enough water and traces of organic matter to nourish colonies of bacteria. The bacteria destroy the plastic, but only slowly, and meanwhile they generate heat, and before long the rubbish dump can start to smoulder rather nastily.

If you can buy PHB (polyhydroxybutyrate) bottles made by Marlborough Biopolymers Ltd., your choices widen. PHB bottles are broken down fairly quickly by soil fungi and bacteria, so you can add them to your garden compost, or just bury them in a flower bed. The bottles are more expensive than other plastic bottles, but once they are in mass production the price should fall. You should not store anything in them for very long, however — they might rot on the shelf!

Waste paper We in Britain could recycle a great deal more of the paper we use, which would save the cost of importing paper and relieve the pressure on the world's forests. Keep paper separate from other rubbish and ask your council whether they can sell paper for recycling, and if not, why not. If that fails, contact Friends of the Earth to see whether there is a voluntary collection, or try writing to Paperback, 8-16 Coronet Street, Hoxton, London N1 6HD for advice.

Booklice are really tropical, but thrive in modern homes and are most common in very new houses. They cannot breed at temperatures below about 20°C, so if you have a nice draughty larder store dry foods there. If not, make sure they are kept in tightly-sealed containers, where not even the tiniest animal can enter. Do not waste money on insecticides. They will contaminate the food and are unlikely to injure the booklice.

Pantry

Wherever food is stored, fungi, bacteria and small animals will try to get in. You can hardly blame them for wanting to eat, but you probably will. How can you protect your food without resorting to poisons?

Booklice If they are stored in a warm damp kitchen, foods such as grains, flour or sugar may be attacked by booklice (usually *Liposcelis bostrychophilus*). These insects, about a millimetre long, are not really lice — they are not parasitic, and their mouths are adapted for chewing rather than piercing the skin of other animals. They transmit no diseases to human beings, but they can distribute the spores of fungi and bacteria which may be harmful, and people tend to reject as contaminated food that is crawling with tiny brown specks.

Canned food Bacteria can be destroyed at temperatures of about 120°C. If clean food is placed in a container, cooked in the container at a high temperature, and then the container is sealed securely while the food is still hot, the food will keep for years without deteriorating. This is called 'canning' because the containers used are usually metal cans, and it was first used in 1810 by the French. Provided the container is sound, the food in it will remain wholesome almost indefinitely.

It is not true that food should be removed from the can as soon as the can is opened. It will deteriorate no faster inside the opened can than outside it.

Canning involves some loss of nutrients, but the loss is much smaller than many people suppose. If canned food is heated to the temperature at which you will eat it, the nutritional content — even of vitamin C — is almost identical to the same food cooked from fresh.

Food storage Food deteriorates in storage because some nutrients are chemically rather unstable, and break down. Vitamin C (ascorbic acid) is the vitamin which is most easily lost. Oils and fats go rancid because their chemical constituents partially break down to form simpler compounds; polyunsaturated fats are more likely to go rancid than saturated ones. Food goes bad because of fungal or bacterial attack leading to its decomposition. Bacteria and fungi need moisture, and they can grow and multiply only within a certain temperature range. They usually reach food in the first place as spores carried in the air, which are present in the air all the time.

The first line of defence against them is to make sure that food is clean and as dry as possible when it is stored, and that unless it is to be used within a few days it is kept in sealed containers. Do not handle food at all if you are suffering from boils or similar skin infections. These are caused by bacteria capable of contaminating food. Good hygiene reduces the risk of contamination.

It is not practicable to keep the pantry at a temperature low enough to prevent bacterial or fungal growth in food which was contaminated before storage, but this is not necessarily a disadvantage. Most bacteria grow best at a temperature between 50 and 60°C, and produce poisons that may remain during cooking at a temperature high enough to kill the bacteria themselves (most bacteria are destroyed if a temperature of about 120°C is maintained for about ten minutes). It is these poisons that cause food poisoning. *Clostridium botulinum* produces one of the deadliest poisons known; it causes botulism, which is often fatal (*see* nitrosamines, page 26). When food is chilled, the low temperature will prevent the further growth of bacteria but will not kill them. As the food is warmed they will proliferate, so the food can be poisoned. If the storage temperature does not prevent their growth, decomposition will be evident

from the smell. The food will be wasted, but at least it will poison no one.

When meat is 'hung' its smell often grows stronger. This is not due to bacterial decomposition, but to chemical changes brought about as enzymes within the meat digest some of the protein — it does not make the food dangerous to eat.

Fungi are less dangerous than bacteria, but cannot be destroyed by low temperatures. They survive well in freezers. They usually become visible when they produce their fruiting bodies (*see* mould, page 37).

Sprouting potatoes If you store potatoes for sowing you will want them to sprout. Store them for long enough in the dark and most potatoes will sprout anyway. On no account should you eat the sprouts — they are poisonous. If potatoes are exposed to bright light for a few hours, green patches appear on them. These patches too are poisonous, as are the stems, leaves, flowers, fruits and seeds of the potato plant. It's a wonder they are not banned!

Potatoes are highly nutritious, consisting of about 18 per cent carbohydrate (mainly as starch) and 2 per cent high quality protein, as well as minerals, vitamins and fibre. They are the main source of vitamin C in the

British diet. They are 78 per cent water, however, so if you eat them boiled or baked in their jackets they are not fattening. When you fry them, some of the water boils away; this reduces the volume so you tend to eat more, which together with the added fat explains why chips and crisps are so fattening.

Weevils Britain has more than 500 species of weevil. They are small, often brightly-coloured beetles, easy to recognize because the front of the head is extended into a long snout or 'rostrum'. The mouth-parts are at the tip of the rostrum, and the antennae, part-way along it, have 'elbows'. The grain weevil, which infests stored cereal products (and ships' biscuits in all the best nautical adventures) is about half a centimetre long. Sailors are supposed to find them more or less edible. If you prefer not to eat weevils, throw away the contaminated food and vow henceforth never to keep flour and other cereal products in paper bags for longer than a week or two. Weevils can drill their way through paper, but not through more substantial containers.

The stairs and upstairs

Stairs

Never mind your fears about pesticides, nuclear power stations or food additives; the stairs are likely to get you long before any of the supposed risks of modern life. Falling down the stairs is one of the most common forms of injury, so make sure the stairs are well lit and have surfaces that will trip no one.

You may feel that it is a waste of energy to provide strong lighting for the stairs, but the cost is not really large. A hundred-watt bulb must burn for ten hours to use one unit of electricity, or put another way, at present prices you can run a hundred-watt bulb for 1¾ hours for a penny. If the cost of the light still worries you, invest in the new long-life bulbs now being marketed. They fit ordinary light sockets, are guaranteed for 6000 hours of use, and consume a quarter of the electricity of an old-fashioned bulb.

Beware of open-tread stairs. Small children are inclined to go right through the gaps, and some infants are better than others at climbing the stairs on the wrong side.

Bedrooms

The bedroom may seem a private place, but environmentally it is rather important. You spend a good deal of time in it, so the environment of the bedroom could affect your health. Is it dusty or well-ventilated, for example, quiet or noisy?

While we are discussing the bedroom it seems appropriate to begin by considering the clothes you keep there, and the origin of the fibres or other materials from which they are made. What environmental price must be paid for us to wear skins and furs? If we prefer not to wear skins and furs, should we grow cotton on land that might grow food? Do synthetic fibres necessarily imply pollution from chemical factories? How is silk produced? Is it less damaging than synthetic fibres?

Fibres

Cotton The cotton plant is a small shrub, the seeds of which develop wrapped in fibres. For every ton of cotton fibre the plants produce two tons of seeds, which are crushed to produce an oil used in making margarine and a cake used to feed livestock. It is not quite true, therefore, that growing cotton precludes growing food. China is by far the largest cotton producer in the world, and the USA comes second.

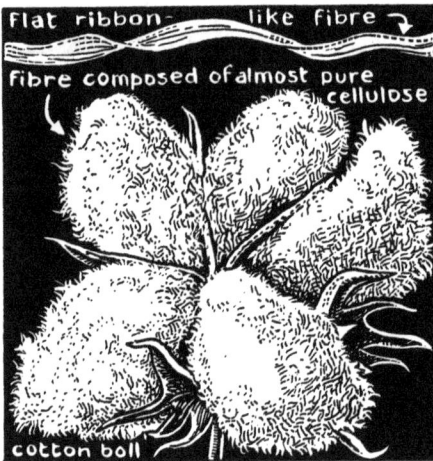

Furs If you were to join an Antarctic expedition you would be kitted out with clothing to withstand the very cold temperatures and strong winds — and it would not be made from furs. If you can survive without furs in the Antarctic, then surely you can in Toronto and New York, never mind Edinburgh. If you like the look of furs, there are artificial imitations which are indistinguishable from the real thing, but do not involve the killing of animals.

Real furs are obtained either by raising suitable animals in captivity or by hunting, and many of the most attractive furs come from rare and protected animals, so the trade is illegal (though given the commercial value of furs the penalties for infringement are mild). Poaching protected species is very tempting in countries where the population is poor, though most of the poaching is in fact carried out by large organizations.

If you want to contribute to stamping out this trade, join Friends of the Earth, Greenpeace, or any of the major wildlife organizations.

Artificial fibres The first artificial fibre was rayon, sometimes called 'artificial silk', which has been made for more than a century. It is manufactured industrially from soft wood, although mulberry leaves, cotton linters and other raw materials have been used. There are several similar processes, but all of them involve treating the raw material with chemicals such as copper sulphate, sulphuric acid, ammonia, or sodium hydroxide (caustic soda) to dissolve the cellulose it contains, and then extruding single filaments.

Synthetic polymers such as nylon, polyesters (sometimes known as 'dacron' or 'terylene') and acrylics are made from organic chemicals derived from petroleum or coal, and treated chemically to produce simple compounds whose single molecules (monomers) can be made to join together in long chains (polymers).

Each of the synthetics has its own characteristics. Nylon is tough and elastic, and is used to make such things as underwear, carpets, and vehicle upholstery.

Polyesters do not absorb water, but can be softened at high temperatures. This allows garments made from them to have permanent pleats and to be crease-resistant.

Acrylics consist of very short flexible fibres, and make light fluffy fabrics used in baby clothes and sweaters.

All the synthetics are produced by the chemical industry, and there is no such thing as a totally leak-proof chemical factory, so their manufacture inevitably involves some pollution, mainly of water. It also consumes energy.

Sheepskin and leather Sheep are raised for meat and wool. When they are slaughtered for meat they may be skinned and the skin cured. If you allow that it is acceptable to raise livestock for meat, then you can accept sheepskins as a useful but incidental by-product. Leather in all its varieties is produced in the same way, and the same argument applies.

The skins of reptiles, on the other hand, are taken mainly from wild animals, and this trade is illegal, although alligators are farmed for their meat and skins in parts of the USA, and trade in farmed alligator skins is legal.

Silk Many arthropods produce single-filament fibres made from protein. Spiders do, and so do the caterpillars of many butterflies and moths. Only one family of moths, the Bombycidae, produces a fibre suitable for making fabrics, however, and most silk is taken from one species within that family, *Bombyx mori*, a moth that has been domesticated for so many thousands of years that it can no longer complete its life cycle without human help.

It is reared under the most intensive and most highly-managed of all factory-farming systems. The larvae feed on mulberry leaves and pupate in cocoons made from the fibres they spin for the purpose. The adult moth leaves its cocoon by dissolving enough fibres to open an exit hole, mates almost at once, and the males are killed as soon as they have mated. Silk from the empty but damaged cocoons consists of short fibres that can be spun to make 'spun silk', but this is regarded as inferior to the finest silk, produced by unreeling a complete cocoon in one long un-

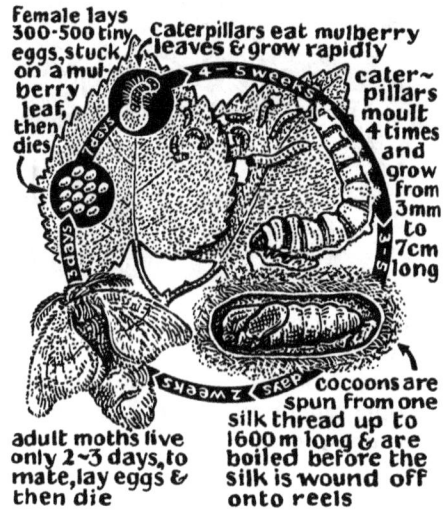

Female lays 300-500 tiny eggs, stuck on a mulberry leaf, then dies

Caterpillars eat mulberry leaves & grow rapidly

caterpillars moult 4 times and grow from 3mm to 7cm long

cocoons are spun from one silk thread up to 1600 m long & are boiled before the silk is wound off onto reels

adult moths live only 2-3 days, to mate, lay eggs & then die

broken filament. In this case the moth cannot be allowed to make its own way out of the cocoon, and is killed, usually by asphyxiation.

Silk is a natural (as opposed to synthetic) fibre, producing very fine, warm, rich textiles, and silk production has no effect whatever on the outside environment. 'Weighted' silk, however, which is produced in small amounts in Europe and Japan, is treated with salts of lead and tin to make it heavier, and with the present concern about the possible toxic effects of even very small doses of lead, you may prefer to check the label to ensure that any silk product you buy is not weighted.

Wool Sheep were domesticated for their milk and meat. It was only some time after domestication that they developed woolly fleeces, and even today sheep are raised mainly for meat, with wool as a secondary product. Wild or semi-wild sheep, such as the Soay, moult each year, shedding their winter coat in straggly bits. Domesticated sheep with fleeces do not moult, and are sheared in late spring when the weather grows warmer and the fleece starts to be uncomfortable. Taking the wool from the animal therefore does no

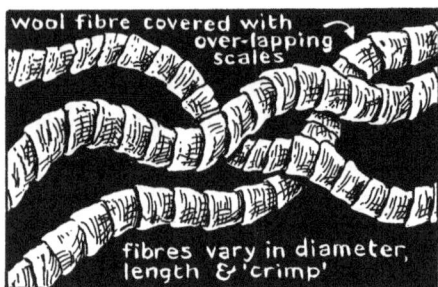

wool fibre covered with over-lapping scales

fibres vary in diameter, length & 'crimp'

more than nature did for its ancestors, and causes no harm whatsoever. Wool is made from animal protein, and some people cannot wear it next to the skin because it causes a rash.

Care of fibres

Clothes moths One stage in the commercial preparation of natural fibres involves treating them with a long-lasting insecticide as a protection against clothes moths. The insecticide usually belongs to the organochloride group. Even the most persistent insecticide fades eventually, however, and if you take care of your clothes they may well outlast their treatment.

There are several species of clothes moth, the most damaging being *Tineola bisselliella*, and all of them produce larvae that can feed on dry plant mat-

tinea pellionella

tineola bisselliella

erial such as cotton, and on keratin, the dry tough material from which feathers and hair are made. The adults do not feed at all, but seek dry warm dark places where they can lay their eggs on a suitable store of larval food.

The best protection is to wash or dry-clean clothes regularly to destroy eggs and kill larvae, keep airing cupboards clean, and make sure the bedding your pets use is kept clean. The moths can establish colonies in a dog or cat basket, or even in a birdcage if there are old feathers lying around — it is but a short trip from there to your Sunday best. Moths can also breed in the nice warm rags used to lag water pipes, so replace these with sponge lagging which you can buy from any builders' merchant.

Fire prevention The fabrics in your house can be grouped into three types: cellulosic, protein, and synthetic polymers. The cellulosic fibres are those of plant origin such as cotton, linen, jute, sisal or hemp, together with artificial fibres of plant origin such as rayon. These will start to char at temperatures above about 150°C, and if the temperature rises higher than that they will burn fiercely. Do not leave them unattended to dry in front of the fire, and do not allow small children to wear flowing garments made from cotton or linen unless the clothes have been treated with a flame retardant.

Other types of fibre are much safer and will not blaze. Protein fibres such as wool, silk, and 'azlon' (which includes the artificial fibres derived from casein, peanuts, maize kernels, soybeans or other plant products) do not burn readily. At about 130°C they will smoulder and melt, but they can be made to burn if they are caught in a fierce fire.

Synthetic polymers melt at high temperatures into a viscous liquid that may then burn, releasing dense choking fumes.

The environment inside the bedroom

Dust All bedrooms generate dust. If you keep the rooms reasonably clean the dust should not trouble you unless you are especially sensitive to it; in this case your response will depend on the degree of your sensitivity. It may be enough just to improve the ventilation and vacuum the rooms more often. Failing that, you could try increasing the humidity (*see* humidity control, page 19), which will encourage the dust to settle. You may need to replace carpets with a smoother floor covering, such as linoleum or vinyl, which is easier to keep dust-free. If all else fails you may have to install dust filters; these are electric fans that circulate the air, blowing it through a filtering device as they do so.

Mites If you are sensitive to dust, it may not be the dust itself that is troubling you so much as the mites in it. The dust consists mostly of flakes of human skin. The outer layer of your skin is made from dead dry cells, which are shed constantly. Every time you rub your hands together you send thousands of them floating off to their doom, and they carry with them vast numbers of microscopic organisms who regard your skin as the entire world. Mites are about the largest of them, and if you doubt that you could possibly harbour them, spend a few minutes examining the skin on your

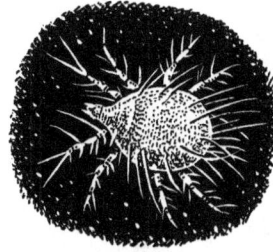

hand or wrist beneath a powerful hand lens. You may see one or two of them as tiny specks that move very slowly. They cause no harm on the skin, but in some people they irritate the linings of respiratory passages.

Noise Few things are more exasperating than the unsilenced scream of the midnight reveller's motorbike. It is an offence to make that much noise, but you cannot complain unless you can take the number of the vehicle, and how are you to do that lying in bed in the dark?

Heavy curtains muffle sound well, and double-glazing — installed for soundproofing rather than for thermal insulation — is fairly effective if you can afford it.

The organization most concerned with this environmental problem is the Noise Abatement Society.

The bathroom and lavatory

The bathroom has the most peculiar environment in the house, as you may discover if you try growing plants in it. In a matter of minutes it can change from being as warm and humid as a tropical rain forest to being as cool and dry as the most arid desert. It is also the place where energy is used prolifically to heat water, and perhaps surprisingly where large numbers of chemicals are used.

Heating

Back-boiler If you burn solid fuel, the fire can function as a central boiler with a closed heating system. The tank in which water is heated is located behind and a little above the fire. It is less efficient than a central boiler because the fuel also has to provide direct space heating, and obviously it works only when the fire is lit, which means it is seldom sufficient by itself. You will need an alternative for use in summer when there is no fire, and you must wait for the fire to heat the water after it has been lit. If your home already heats water in this way you will

have to make the best of it, but if not, fitting a back-boiler is expensive, complicated, and makes no economic sense.

Central boiler A central boiler burning gas or oil is less efficient at heating water than an immersion heater, but it may also provide central heating. Make sure its burners work efficiently and are kept clean, that there is adequate ventilation to prevent the production of carbon monoxide, that its time switches are set to provide heat only when it is needed, and that the thermostat setting is as low as possible. Try moving the setting a little lower each day until somebody complains about a cold bath.

Most boilers use a closed system to heat the water. There is a small water tank heated directly by the flames, connected to a system of pipes that allows the water to rise by convection to the hot water tank, where it enters at the top, flows through a coil of pipes to the bottom, and returns by gravity to the boiler. There is a small header tank at the top of the system, kept full from the cold water tank and used to compensate for any small losses of water. The water in the heating system never comes into direct contact with the water you use.

District heating If they plan to build a big power station or large factory close to where you live you are entitled to object, and someone is supposed to listen to your objection. You could try making a deal whereby you and your neighbours will withdraw your objection provided the power station or factory supplies you with hot water. Ideally the hot water should be free, but at all events it should cost much less than you are paying to heat it any other way. The 'district heating' scheme may not be practicable for technical reasons, but if it is it will feed cooling water from the plant through a series of pipes that leads through all the houses, warming each hot water tank as it passes (*see also* combined heat and power, page 11).

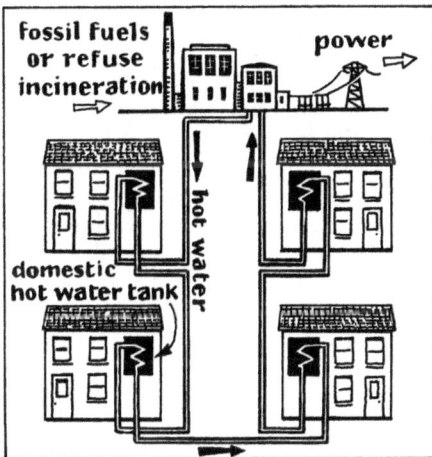

Electric heaters When you run the hot water, some of it will evaporate and condense on the walls, trickling down as rivulets so small you may not notice them. If you have any object fixed to the wall, water may lodge against it, seeking any tiny opening into which it might flow. If the object happens to be an inadequately-earthed electric heater the consequences can be lethal. Water is a good conductor of electricity and will provide it with a route to the earth. If you must have an electric heater in the bathroom, make sure it is fixed high on a wall, well out of the reach even of people waving their arms and towels about as they dry themselves, and make sure it is fitted by a competent electrician who will ensure that it is properly earthed.

On no account allow an electric fire or heater to stand on the bathroom floor while people are bathing or showering.

Gas water heaters The old-fashioned geyser, fastened to the bathroom or kitchen wall, consists of a coiled water pipe and a gas burner. Cold water enters at the top, flows down through the coiled pipe to the tap at the bottom, and is heated by gas flames as it flows. The heater works well provided the coil is long enough for the water to be heated adequately before it reaches the tap.

It is efficient because it burns fuel only when hot water is needed, and the hot water is used at once, so it does not have to be raised to a temperature higher than is necessary to allow for cooling in a tank.

There is a serious disadvantage, however. Unless the supply of oxygen to the burners is sufficient to support proper combustion of the gas, the heater will only partly oxidize its fuel. It will produce some carbon monoxide, which is poisonous. The danger can be avoided only by ensuring thorough ventilation, so that the hot waste gas (which should consist almost wholly of carbon dioxide and hot air) escapes upwards into cooler air, where it can continue to rise. This creates a draught to draw a steady stream of air into the bottom of the heater. The waste gas must be carried to the outside of the house, so you cannot simply buy a gas water heater, fix it to the wall and start using it. You must have it installed by a qualified fitter, and if you have an old geyser, have it checked to see that it is working safely and efficiently. It may

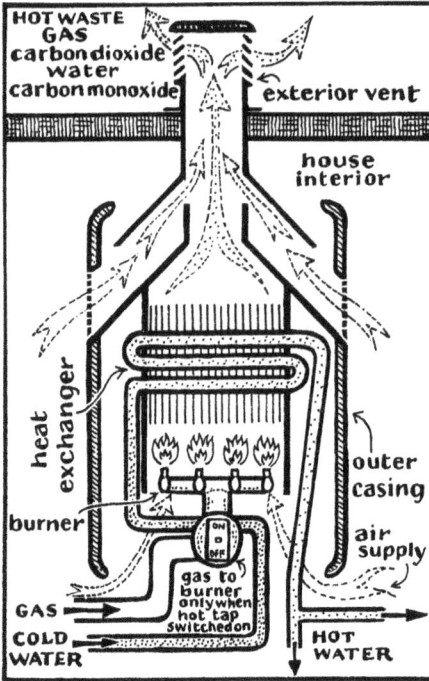

HOT WASTE
GAS
carbon dioxide
water
carbon monoxide

exterior vent

house
interior

heat exchanger

burner

outer
casing

air
supply

ON O OFF

gas to
burner
only when
hot tap
switched on

GAS

COLD
WATER

HOT
WATER

not have been linked to a good ventilation system, or an old ventilating brick may have been blocked or covered, and one bath night 'old faithful' may kill somebody.

Electric water heaters resemble gas heaters, but require no special ventilation because they burn no fuel.

Immersion heaters If you fit an immersion heater to your hot water tank it becomes a kind of big electric kettle. It is efficient because the heating element is immersed in the water that is to be heated, so heat is not lost to any surrounding air — or at least, not from the heating element itself. It will however, be lost from the tank itself unless it is wrapped in a jacket thick enough to feel cool on the outside even when the water in the tank is at its hottest. Do not overheat the water; as its temperature rises the amount of energy needed to raise it by a further degree increases rapidly. Set the thermostat to 55°C, which is as hot as you really need it, and fit a time switch so the water is hot

when it is needed but not being heated when everyone is out of the house or asleep.

Showers or tubs? It takes much less fuel to heat the water for a shower than to heat enough water to fill a bathtub, so showers save energy and are cheaper, provided you remain under the shower for just three or four minutes, which is plenty of time to get clean. If you spend longer, using a constant flow of water, the shower becomes more expensive than a tub as soon as a tub's worth of water has gone down the drain. Keep it brief.

The bathroom environment

Damp The first evidence you have that the bathroom is too damp may be paint or wallpaper peeling from the walls or ceiling, or black patches on the walls, the fruiting bodies of very small fungi. Behind the scenes plaster and woodwork may also be damp, causing plaster to crumble and providing in the wood conditions suitable for dry rot or damp rot. It can also penetrate old and damaged electrical insulation, allowing electricity to earth itself by taking a

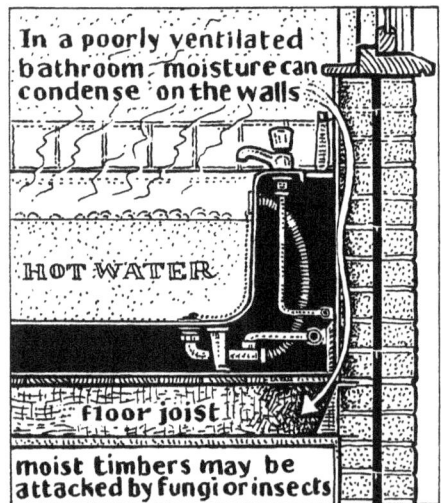

In a poorly ventilated bathroom, moisture can condense on the walls

HOT WATER

floor joist

moist timbers may be attacked by fungi or insects

short cut, and generating heat that may eventually cause a fire.

You should not leave damp untreated or the cause of it unremedied, but you should not have to resort to fungicides. Deal first with the cause, which is most likely to be poor ventilation. Improving the ventilation may necessitate structural alterations, but opening a window will help.

Strip off affected paper and paint, remove and replace damaged plaster, and apply a sealant before redecorating. This will be a paint-like water-resistant substance (make sure there is plenty of ventilation while you apply it), or a special paper, foil or polystyrene covering. Its purpose is to protect your decorations from dampness remaining in the wall.

Ultraviolet (UV) radiation The Sun emits radiation over a wide waveband. At wavelengths of 4 to 400 nm (1 nanometre = one thousand-millionth of a metre), the radiation is known as 'ultraviolet' (UV). Most of the Sun's UV radiation is absorbed by ozone in the upper atmosphere. Only that between 300 and 400nm reaches the surface. UV radiation at much less than 300nm is damaging to some living cells. UV lamps emit only long-wave UV. Glass absorbs UV radiation, so ordinary windows provide total protection.

Sun (UV) lamps When human skin is exposed to UV radiation certain chemicals undergo a change, producing vitamin D_3. A short daily walk in ordinary daylight provides sufficient exposure for a human to manufacture adequate amounts of the vitamin.

The skin also protects itself from damage by increasing its output of melanin, a dark-coloured substance produced in specialized cells, which absorbs UV. A sun-tan is therefore a sign that the body has protected itself from injury. It may be only partially effective. Prolonged sun-bathing can cause a mild, non-malignant form of skin cancer.

A sun-lamp consists of a tube containing some argon gas and a small amount of mercury. A small electric arc across electrodes in the tubes ionizes some atoms of argon, the ionized atoms diffuse through the tube and cause a main arc, the main arc vaporizes some of the mercury, and the ionized mercury atoms emit bluish-violet light and UV. The outer layer of the tube filters out excess UV and UV whose wavelength is short enough to damage cells. A sun-lamp does not make you healthier and some scientists say that all sun-bathing is dangerous, including basking beneath a UV lamp (*see also* sunbathing, page 14).

Chemicals

Bleach The bleach you use in the lavatory and bathroom is exactly the same as the one you use in the kitchen, and just as unnecessary (*see* page 31).

Lavatory disinfectants Living organisms present in human excrement cannot survive long outside the human body, many are harmless, they are removed efficiently by flushing in a flush toilet, and provided you observe ordinary rules of personal hygiene the lavatory presents no risk to health. Disinfectants are unnecessary.

Most domestic disinfectants are made from sodium hypochlorite in various concentrations. It releases chlorine, which gives it a 'clean' smell. Such disinfectants are more effective than ordinary soap and detergent, but not much more, and are deactivated at once on contact with proteins. They will mask a bad smell by producing a stronger one of their own. You could do as well by opening a window.

Sodium hypochlorite is harmful to humans if swallowed, splashed into the eyes, or on wet skin, but it causes no pollution in the sewerage system.

Medicines We all have old medicines and potions sitting on bathroom shelves in boxes and dark bottles. It is a mistake to keep medicines — they can cause accidents, and most of them change chemically after a time. You should dispose of medicines as soon as you stop taking them.

Do not flush them down the toilet or pour them down the sink, however, unless you are quite sure they are innocuous. They can pollute water and may be very potent. Take them to a dispensing chemist, who will dispose of them safely.

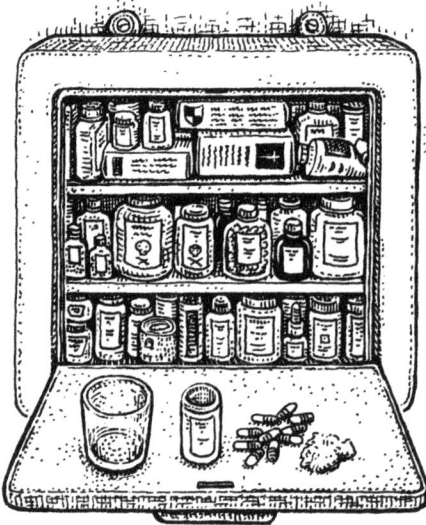

Nickel Most metallic alloys, and especially stainless steel, contain some nickel, and supposedly pure metals often contain it as a contaminant. In some people it causes eczema, and as with all such reactions, sensitivity increases with exposure until even the edge of the buckle of a leather watch-strap can produce a rash. If you develop a rash, replace the strap with one that has no metal against the skin.

Shampoos People used to wash their hair with ordinary soap. Provided you use enough soap to produce a thick lather, soap has one advantage and one disadvantage. First the good news: it is not highly efficient, so it cleans the hair and scalp without removing all the natural oils from the hair. This leaves the hair supple and in good condition. The bad news is that it reacts with the mineral salts in hard water to form a scum that coats the hair and scalp and is difficult to remove, although rinsing in a mild acid such as lemon juice or vinegar helps.

Modern shampoos are mild detergents, and do not form scum. Their disadvantage is that they may remove too much of the natural oils, leaving the hair brittle, dry and lustreless. Like all domestic detergents, shampoos are biodegradable and do not cause serious pollution of water. Use them sparingly, however, to protect your hair rather than the environment.

Technology

Dry toilets A few years ago there were many articles about toilets you did not flush. They used no water at all, so they wasted no water. Some people actually bought them. Provided you have a house with more than two storeys, the theory is very simple.

The bowl is shaped like a funnel, wider at the bottom than the top, so material falls through it without touching the sides, and the top fits snugly, so the bowl is sealed except when it is being used. The material falls through a sloping container partly filled with ready-made garden compost, to which vegetable or garden scraps are added from time to time through a snug-fitting access cover.

The mixture makes a compost, heating as it does so. The warm air from the compost heap leaves through a ventilation pipe and is replaced by air drawn down from the bowl above it so material is disposed of efficiently, and odours are removed by drawing them downwards in the natural airflow.

The lavatories were invented in Scandinavia and used mainly in holiday houses with no piped water supply. They were not cheap, but undoubtedly they were pleasanter than the chemical toilets they were meant to replace. They did not work as efficiently as they should, and are not appropriate for British homes which do have piped water.

At about the time they were attracting attention, some people were busy saving water by partly filling their lavatory cisterns with bricks. The bricks reduced the amount of water the cistern could hold, and thus the amount of water used at each flushing. This saved water but it also increased pollution, because unless sewage is treated and more or less pure water is discharged into rivers or the sea, dilution of the pollutants is very important. The more diluted the pollutants, the less harm they cause, and flushing the lavatory dilutes its

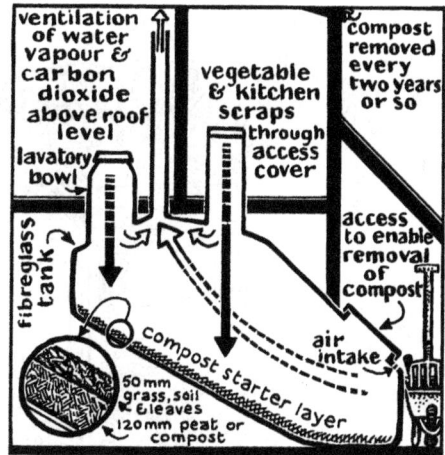

contents as it removes them. If you reduce the flow of water you decrease the amount of dilution, discharge a more concentrated effluent, and can increase pollution, especially during dry weather when the natural flow of rivers is reduced.

Even the water-saving theory was based on a misconception. Britain is not short of water, but it used to be short of reservoirs in which to store it. That is not the same thing.

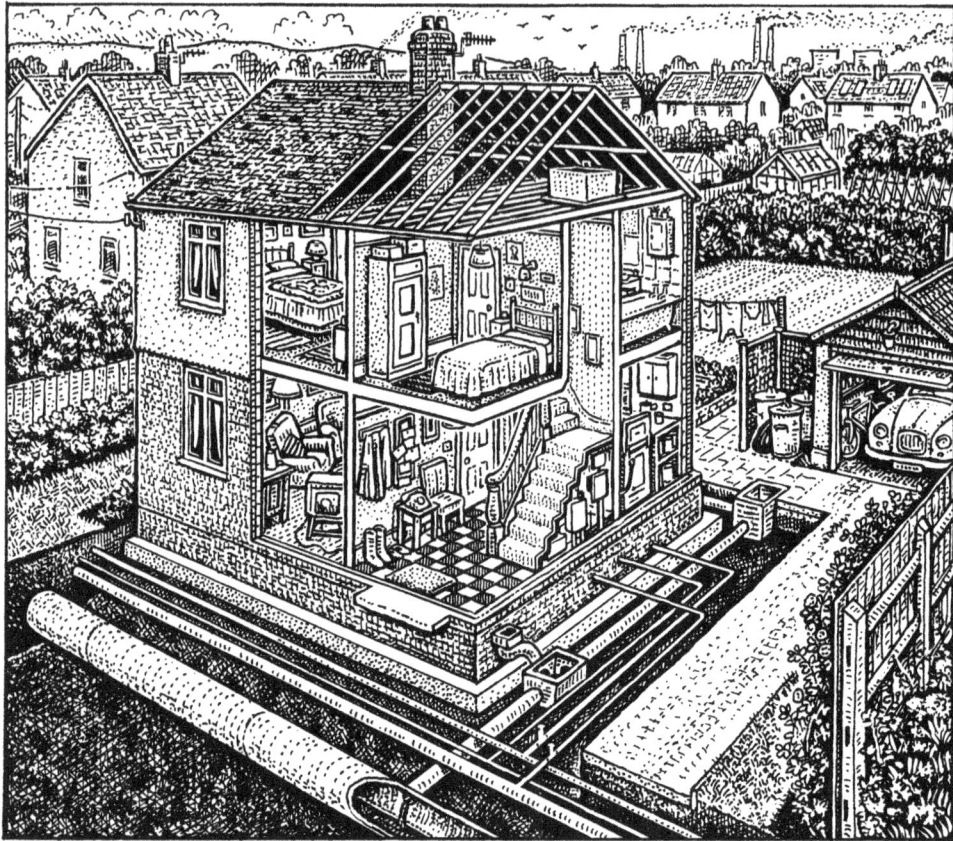

The house itself

The roof

If you leave the unoccupied space beneath the roof uninsulated, expensively-warmed air will heat it pointlessly and heat will be lost through the roof itself into the outside air. If you insulate the space efficiently you may make it cold and damp, and encourage the growth of fungi. What should you do for the best?

Should you think of using the roof itself as a source of energy, by installing devices to trap solar energy?

Pests

Death watch beetle A relative of the woodworm or furniture beetle, the death watch beetle (*Xestobium rufovillosum*) works in the same way. The female lays eggs on small crevices in a wooden surface, the larvae eat their way into the wood making galleries, and the adult beetles leave through small exit holes they make for the purpose. You may see the exit holes, but if not the first you may know about the infestation is when the timber collapses in a cloud of dust. This can be most embarrassing if the timber in question is holding up your roof.

You may hear the death watch beetle, however, because the adults have very hard heads, which they knock vigorously in short rapid bursts against the walls of their galleries. The knocking is believed to be a form of communication, perhaps concerned with mating.

The beetles prefer soft, damp wood. They will not attack wood that is fully matured and hard, very smooth, or covered with a good coat of paint or varnish. Vulnerable wood can be protected by insecticide, applied by a specialist contractor. Used in the roof space or other out-of-the-way places, the insecticide will not harm you, your

death watch beetle

adult taps its head 7 or 8 times a second on the timber

larva

larva

wood debris left by larvae

pets, or wildlife, but you must not use insecticides in a roof used by roosting bats (*see* bats, page 109). The presence of bats does not increase the likelihood of beetle attack because the bats will devour any beetle that comes anywhere near them.

Dry rot Dry rot leaves timber apparently dry, and characteristically broken into small rectangular pieces. Structurally, such wood is useless and often dangerous. The wood may look dry, but the rot can only penetrate where the wood is damp.

The trouble is caused by *Serpula lacrymans*, a common fungus whose spores are present in the air all the time. When spores land in a suitable environment the fungus starts to grow as a mycelium, a network of very fine thread-like hyphae which absorb nutrients. It is the mycelium that breaks the wood into very small pieces, and it can also penetrate plaster, brickwork, and even concrete. When it reproduces, the fungus produces distinct patches of fleshy brown fruiting bodies, but by then the damage has been done.

If your home is attacked by dry rot you need expert help. All affected wood and plaster will have to be removed and destroyed, fungicides will have to be sprayed everywhere, and it will be expensive. The best protection

dry rot fungus
millions of spores released every minute

is to make sure that your home is kept dry. This applies throughout the house, but you should pay special attention to the roof timbers, especially if the loft is well insulated, because then it may be cold and damp. If in doubt, paint the inside timbers with a wood preservative.

Saving energy

Loft insulation Warm air rises, so it is not surprising that more heat leaves a building through the ceilings and roof than through the walls and windows. If you have not yet insulated your loft or roof space with fibre, do it now, and lay it at least ten centimetres thick. Remember not to insulate below the cold water tank, which could freeze if you do.

Eave to ridge ventilation helps prevent moisture, rising through the insulation, condensing in the cooler air of the attic. 10cm of insulation material between loft joists, not sealing the eaves, but taken over cold water tank, and access trap door

water tank

Warm air, rising through the house, is trapped below the loft insulation

When the job is finished, the roof space will be separated from the rest of the house, and in winter it will be very much colder, though sheltered from drying winds. This may encourage water to condense on to exposed timbers, and condensation may drip on to timbers from the underside of roof slates. From time to time you should check that the roof space and roof timbers are not unduly damp. If they are, you will have to leave the access door open long enough to dry out the area, and you may need to treat the timbers to guard against rot.

You can exploit the temperature difference, however, by installing a heat pump in the roof space. It will cool the area around it still more, but deliver heat it extracts to the house below. It is very efficient and mechanically simple (*see* heat pumps, page 12).

Mineral fibres The traditional mineral fibre, used mainly for fireproofing, but also for roofs and lagging, was asbestos. There are several types of asbestos, but nowadays all of them are banned for most uses, because minute fibres, detached when the asbestos is worked but especially when it is cut, can be inhaled and lodge in the lungs, causing illness. Other mineral fibres are used instead, and as far as anyone knows these are perfectly safe to use.

Water tanks In British houses the cold water tank is usually located in the roof space. When you insulate the roof space, avoiding the water tank, remember to lag all the pipes entering and leaving the tank very thoroughly. It will be very cold in the roof space in winter, and without lagging the pipes will certainly freeze.

Unless you are a plumber it is not worth altering the location of the tank in an existing house, but if you are moving to a house that is still being built, try suggesting one of the alternative arrangements found in most other high-latitude countries. Either install the cold water tank inside the house rather than in the roof space, if necessary using a small electric pump to push the water to upstairs taps and the hot water tank, or dispense with a cold water tank entirely, which is the usual method. The cold taps, lavatory cistern and hot water tank are fed directly from the mains. Once they come indoors all pipes are kept inside the house, where they are unlikely to

freeze, but further protection is given by the pressure of water in them, which is usually enough to dislodge ice before it blocks them completely.

Generating energy

Solar cells Sunshine is warm only in summer, and if you live in a high latitude summer is short. Sunlight is available every day of the year, and in high latitudes summer days are very long. If you could exploit the light of the Sun rather than its heat, solar energy would be more abundant and could be supplied throughout the year.

Solar cells were developed to provide the power needed to operate the instruments and controls on space satellites. They convert sunlight into electricity. As with the solar panel (*see* below), the principle is simple. Place two wafers cut from a crystal of silicon next to one another and add impurities to change the configuration of electrons within them. When light strikes them the electrons will move across the junction from one wafer to the other, and you have a solar cell. It is not practicable to make such cells very large, so an array of cells is required for the generation of a useful current.

At present solar cells are expensive and not very efficient, so they are of limited use, but scientists are working hard to improve them and their price is falling. In years to come it may make sense to cover your roof with them and let the Sun generate some of the electricity you use.

Solar panels The best things in life are free, and what could be better than the warmth of the Sun? To some extent you use it already since it warms your house in summer; houses should always be built so that one large wall faces the hottest position of the Sun (in Britain this is within about 30° of south, but remember that true south is about 6° west of south as indicated

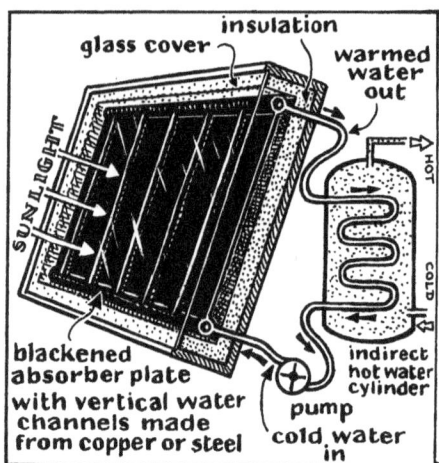

glass cover insulation
warmed water out
SUNLIGHT
HOT
COLD
blackened absorber plate with vertical water channels made from copper or steel
indirect hot water cylinder
pump
cold water in

by a magnetic compass).

You can go further and use it to heat water, a technique used quite extensively to heat swimming pools. The trouble is that although the warmth is free, the equipment you need to transfer it to the hot water system is expensive. In Britain it is cheaper to heat your water by electricity or gas than to fit solar panels to an existing house, but if you are building a new house and can include the panels and plumbing as part of the design, the economics may work out more favourably. If you live in the south of England you should certainly consider it.

The principle is very simple. Panels (at an angle of about 35° to the horizontal) on the roof or wall that face the Sun are coated with a matt black finish so they absorb heat. Inside, the panels contain piping through which the water passes slowly enough to be heated. The pipes then carry the water to a heat exchanger, another set of coiled pipes inside the hot water tank, and from there back to the panels. You need a small electric pump to circulate the water, a header tank to compensate for inevitable small losses of water within the circulating system, and heat sensors to control the switch that turns on the pump. The sensors are necessary because if the water circulates when the

temperature in the hot water tank is higher than the temperature in the panels, the hot water will be cooled rather than warmed.

If you are good at making things and understand plumbing, you should be able to design, make and install your own solar heating system, but if not then contact one of the firms listed in *Yellow Pages*. Remember that the system will work best on hot days in summer, but hardly at all in winter, so do not try to make it heat radiators.

The exterior

Lead and zinc flashing If you look at the outside of the roof of your house you will notice strips of metal around the places where chimneys and dormer windows join the roof. This is 'flashing', which allows rainwater to flow past and away from the join rather than through it to the inside of the house, and if there is a crack or break in the flashing you will have a leaky roof followed by a repair bill.

The flashing is made from lead, or sometimes zinc. Rainwater, which is slightly acid, will react with the metal, so the water in the gutters will contain small amounts of lead or zinc compounds. The level of pollution is not serious provided the rain completes its journey to the ground, where it can be diluted in a much larger volume that has not crossed roofs. You should not drink rainwater you have collected from the roof, however, because of the metals it may contain.

Thatch A thatched roof provides excellent thermal insulation for the building beneath, but it is also a habitat for many small animals. Made from reed, a thatch should last for eighty years before it needs renewing; a straw thatch does not last so long. Thatch is much less of a fire hazard than it seems. It can be treated to make it partly

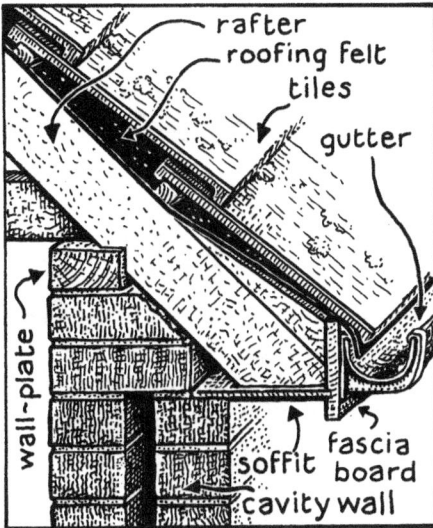

Labels in illustration: rafter, roofing felt, tiles, gutter, wall-plate, soffit, fascia board, cavity wall

fireproof, it is often damp from rain, and in any case close-packed material does not burn well.

Eaves The roof of your house has eaves to carry rainwater clear of the walls and into the projecting gutter. If a gutter is blocked or broken you should clear or mend it quickly, because leaking water can be blown back on to the wall, which it will penetrate to cause damp. At the top of the wall there is usually a wooden fascia board.

Beneath the eaves there is an area sheltered from the rain and to some extent from the wind, available to any plant or animal that can manage not to fall off. House martins can fix their nests of mud plaster directly to the fascia boards, and smaller birds can nest in the tiny space between the top of the fascia board and the roof slates. There are often gaps between the fascia boards and the slates, providing access to the roof space inside the house to any animal small enough to pass through them. Respect these lodgers. Try not to disturb them when you make repairs, and if you have nesting birds do not paint the fascia boards during the nesting season in spring and early summer. Wait until the young

have left the nest.

Do not remove moss that may grow on the roof itself. Birds will do that for you and obtain a meal as they do so. The moss conceals a small reservoir of organic debris which is food for very small insects. Insect-eating birds hunt among the moss, ripping it open to probe below the surface, often dislodging much of it as they do so.

Walls

How polluting are the industrial processes by which stones, sands and clays are converted into building materials? How can you modify your walls to make them retain heat and exclude water more effectively?

Block Building blocks are made from concrete, or from coal cinders, sand and cement to make a lighter 'breeze block'. Concrete blocks are satisfactory to use, though they are heavier and provide poorer thermal insulation than brick. Breeze blocks are much lighter, provide better insulation than concrete, and also absorb sound better. They are partly made from wastes (cinders and small scraps of stone) and so have environmental advantages, while their rough surface provides good bonding for mortar and exterior rendering. *See also* concrete, page 62.

Brick Brick can look beautiful, especially when it has had a few years to weather to its proper dark rich colour. The vicinity of a brickfield is not always a pleasant place to live, however. Bricks are made from clay, these days dug from deep wide pits which remain as industrial wasteland. The clay is ground and screened, then either mixed with water and moulded, or passed dry into a press and stamped into shape. The 'green' bricks are then dried and kilned at a temperature high enough to vitrify the clay, so the finished brick is a ceramic product.

The kilning uses much energy, and releases gases that smell offensive and include sulphur dioxide and fluorides. In the past these have caused serious pollution. Sulphur dioxide is corrosive, and fluorides adhere readily to grass in fields downwind of the kiln. They make the grass grow lush, but grazing cattle suffer an excessive growth of bone, leading to deformities and a softening of the teeth. Like all pollution problems, those associated with brickfields can be remedied, but it is important to remember that there is an environmental price to be paid even for such a traditional and apparently innocuous article as a housebrick.

Concrete Concrete is a mixture of cement, gravel and sand, easy to store, and clean and safe to use around the home and garden. If used carefully it is strong and durable. It causes no environmental problems while it is being used, and gravel and sand are obtained by ordinary quarrying, but have you ever explored the area around a cement works?

Cement begins as limestone. The rock is quarried, then crushed and ground to a fine powder that blows into the air and coats everything for miles around. The powder is blended with other ingredients, usually clay, sand, and some iron ore, and then burned in a kiln at a high temperature. The kilning drives off carbon dioxide (*see* greenhouse effect, page 14) from the limestone, leaving quicklime (calcium oxide), and causes chemical reactions among the various ingredients. After kilning the cement is ground again. The process is noisy, dirty, and uses a great deal of fuel. Use concrete sparingly, and have a thought for the greenhouse effect and the people who live downwind of the cement works.

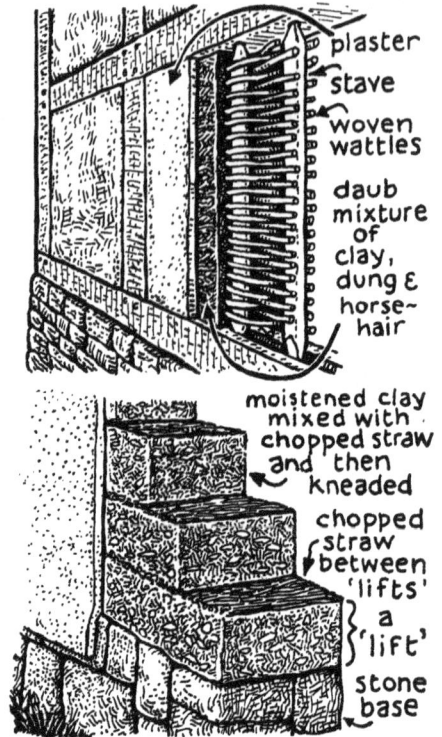

Mud People have traditionally built their houses from whatever materials lay close to hand. In the forests of Russia, Scandinavia and North America they used timber, on windswept islands they used stone, and where there was suitable clay they used brick.

The most plentiful of all materials, however, is soil, and it too can be used. There are large and substantial daub-and-wattle houses made from mud mixed with straw that have stood for centuries. You can see them all over East Anglia. Cob, a West Country variant made from mud and small stones, can be almost as durable.

If you have a chance to buy such a house, do not let the nature of the construction deter you, but make sure you keep the walls in good repair and well covered with waterproof paint. Mud walls have been known to wash away!

There is a modern version of this traditional method of using soil for building, and you may like to consider it if you are planning a new house. It is a little more complicated than plastering the material on to a wooden framework. Stone or timber is used to build a 'skeleton' that bears the weight of the roof and floors, then the spaces are filled with 'rammed earth'. A movable frame is made from metal or stout timber, with a space between the timbers the thickness of the wall. Earth dug from well below the topsoil is packed into the frame, rammed down hard with a heavy weight, then the frame moved on to the next section to be built. Finally the walls are coated on the outside with a weatherproofing material.

This building material is obtained locally and causes no environmental damage whatever. The hole excavated during building can be turned into an attractive pond or swimming pool.

Plaster If you heat gypsum (calcium sulphate), let it cool, grind it to a fine powder and mix it with water, you have 'plaster of Paris'. Most plasters are made from plaster of Paris mixed with lime, sand and cement in varying proportions depending on how it is to be used. Modern plasters used to fill holes and cracks in existing plaster are usually made from cement or cellulose mixed with a synthetic resin. All these plasters are harmless and non-toxic, and their manufacture causes no serious environmental problems. *See also* concrete, page 62.

Stone Stone is probably the best of all building materials, though it is expensive. It is obtained by quarrying, which damages the environment, but the damage is of a traditional kind and does have advantages. As the stone is cut, rock faces are exposed that are always of interest to geologists, and sometimes they are of major scientific importance

in helping us to understand the Earth and its history. An abandoned quarry soon loses its raw appearance as it is colonized by plants and animals, and in time a former quarry may become a small area of wilderness, supporting a variety of wildlife. Most rocks contain a little radium and emit radon, but this is seldom a problem (*see* radon, page 20).

Wood Every house contains timber, in joists, beams, floors and doors, and some houses are constructed from little else. Most of our furniture is made from wood, our shelves are wooden, and although they call this the 'atomic age' or the 'computer age', most of what you see if you look around is still wooden — in many ways we are still living in a 'wooden age'.

There are two ways to obtain wood. Either we grow it in our own forests, or we import it from forests in other parts of the world. Britain has more trees today than it has had for a thousand years, and the foresters want to grow

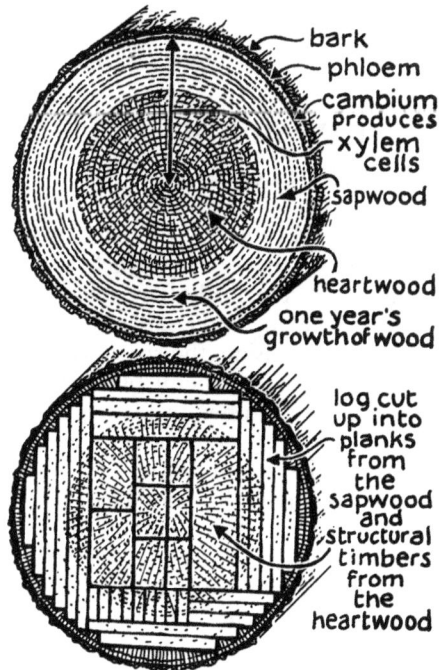

bark
phloem
cambium produces
xylem cells
sapwood
heartwood
one year's growth of wood

log cut up into planks from the sapwood and structural timbers from the heartwood

more, but people complain. The dark conifers march in serried ranks across open moorlands and high hills, and the prospect of still more afforestation, perhaps encroaching on the lowlands too, fills conservationists with alarm.

So perhaps we should import more timber. We could do so, but there are fears about the deforestation of the tropics, caused partly by the felling of trees to fulfil our demand for them.

Natural wood is beautiful, easy to work and familiar, but environmentally it is not cheap. Use it sparingly, but use alternatives if possible. Use chipboard wherever possible, because this is made from the whole tree, including the smallest twigs that would otherwise be wasted, chopped small and compressed. The production of chipboard uses forests more economically by clearing an area completely, using all the material, and allowing the clearing to be replanted. The alternative is to take isolated trees from a much larger area, damaging many more in the process of felling and extraction, or to clear a large area, in either case wasting most of the wood because only the trunks and larger branches are used.

Energy-saving and protection

Cavity insulation Old houses, made from stone, have very thick walls. Modern houses usually have thin walls, with brick as the outer skin and breeze blocks as the inside, the space between being filled with mineral fibre, plastic foam, or some other insulating material. There are some houses with a cavity wall construction, however, where the cavity has not been filled. The cavity can be filled by drilling holes through the outer skin at strategic points and injecting loose mineral fibres to fill the space, or urea formaldehyde foam which fills the space and then sets hard. The operation is per-

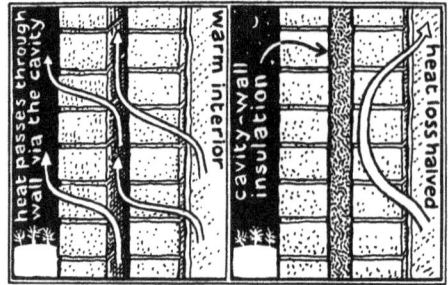

formed from the outside, so it brings nothing into the house, but there can be risks, and you should consult a reputable contractor before allowing any work to begin.

Moisture cannot easily cross an air space, but it can move by capillarity through the spaces in fibres or a foam, and if this should happen then both the skins of the wall will be made damp and the problem will be extremely difficult to remedy.

You can insulate your walls without filling the cavity. You can hang heavy curtains to the inside of external walls, of fix a thick layer of expanded polystyrene to them. You can coat the outside of the walls with glass fibre, weatherboarding, or some other insulating material. The effect will be as good as filling the cavity, with no risk of penetrating damp.

Damp-proofing Modern houses are built with a damp-proof course, which is a layer of impermeable material built into the outside walls just above ground level. Houses built before the early years of this century had no damp-proof course, and old courses, sometimes made from slate, may break. Damp can then rise up the walls from the ground.

The remedy is not difficult and does not involve the use of hazardous chemicals. A silicone, which is a polymer compound based on silicon and oxygen, can be injected through holes drilled into the base of the wall. The silicone penetrates brick and makes it

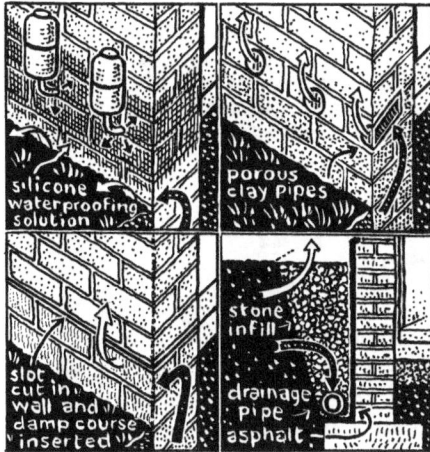

waterproof.

Alternatively the wall may be drained by inserting small cylinders at intervals which draw water and allow it to drip away harmlessly.

The most elaborate method is 'tanking', where the soil is stripped away to the foundations and the outside walls treated with an impermeable sealant.

Foundations

There is not much you can do safely to alter the foundations on which your house stands, but if you are planning a new house you might use the foundations to reduce your heating bills.

Insulation When a modern house is built, the foundations are dug deeper than 90 centimetres or so only if the ground is soft. If the foundations are dug to twice this depth, the outer walls fully insulated, and the space enclosed by the walls filled with earth to the usual foundation depth, then the finished house stands on a block of earth held by its outside walls. This block acts as a heat store. When the house is warm, heat passes through the ground floor to warm the earth beneath it, where it is trapped by the insulated walls. When the house is cool it is warmed by the earth it stands on.

Drains

What goes down the drains, and what effect does it have down there, out of sight and out of mind? Will the chemicals you use to get rid of a nasty smell kill a harmless fish many miles away?

Disinfectants If there is a smell from an outside drain it is because material is trapped down there and is decomposing noisomely. A strong disinfectant, which is extremely poisonous to humans and all wildlife, will arrest the decomposition and simply impose a stronger smell to mask the unpleasant one. It will not solve the problem; to do that you need to free the trapped material. Try flushing the obstruction clear with clean water under pressure, perhaps by fixing a garden hose to the nearest tap and pointing the end down the drain. If that fails, try fishing down the drain with a bent wire. If that is not sufficient you may need to call a plumber. At no stage do you need disinfectants.

Sewage If your house is connected to a mains sewage system your wastes will

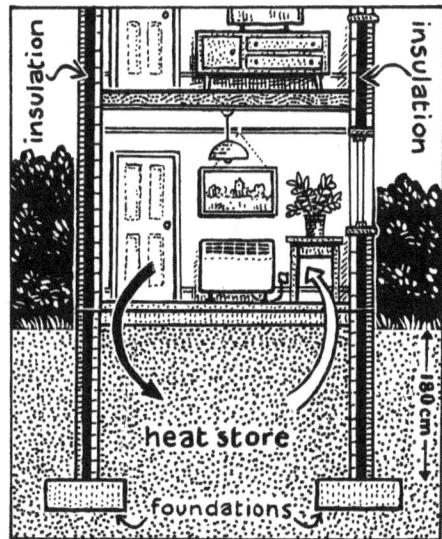

SEPTIC TANK & BIOLOGICAL FILTER

dry crusty scum formed on the surface from lighter solids rising through the liquid

organic matter in the effluent oxidised by aerobic bacteria in a surface film on the broken stone in the filter

methane from decomposition vented off

sewage

cover

scum

sludge

rough broken stone or clinker

raw sewage undergoes anaerobic decomposition by bacteria and fungi with a sludge being deposited

effluent outfall

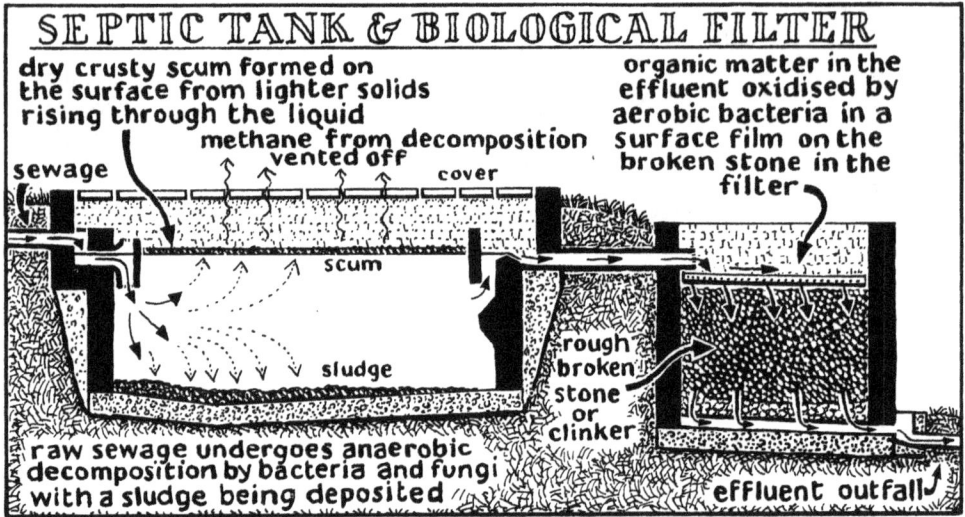

be spirited away, bidding you farewell with only a parting gurgle. You may think that is the end of them, but in the real world everything has to go somewhere. Sewage should go to a treatment works where it is screened to separate out the solids, which should be dried to make a sludge which can be used as a fuel or a fertilizer, while the liquid should be purified before being discharged into a river or the sea. It is just as likely, however, that it will receive little or no treatment before being discharged into the receiving waters. Why not find out what happens to your sewage, and if sewage treatment is inadequate in your area, start campaigning? After all, you could be on the receiving end.

If you do not have mains sewerage you will either use a cesspit or a septic tank. A cesspit is an impermeable container buried somewhere in your garden. When it is full a tanker will come and empty it. A septic tank is also a container, but with permeable sides, and it contains bacteria to break down organic waste. Water seeps from the sides, which act as a very fine filter, and is purified by flowing down through the soil to the groundwater. The solids decompose in the tank. A septic tank needs emptying only very rarely, and is a highly efficient way to dispose of wastes.

Outside the house

Garage and sheds

You may not feel up to building your own house, but you might try your hand at a garage or shed. When you do you will face a bewildering selection of techniques and materials. How safe are they, both for you and for the birds and other animals that may visit them? Once the buildings are finished you will store things in them, and you may think of using a shed to house an electric generator. Is this a good idea?

Materials

Asbestos roofs Asbestos sheeting makes an excellent roofing material for outbuildings. If there is nothing wrong with an asbestos roof, leave it alone; it is safest where it is. Do not attempt to remove or alter an asbestos roof by yourself. Asbestos sheeting is harmless, but loose fibres can harm you if you inhale them, and you will release fibres if you drill, saw or break the sheeting. In Britain it is illegal to work asbestos (and this includes removing old asbestos) unless you are licensed to do so,

use the proper equipment and protective clothing, and make sure that special arrangements are made for the final disposal of the asbestos. Look in *Yellow Pages* for the names of authorized contractors.

Plastic roofs Modern materials, based on rigid or semi-rigid polymers, are strong, in many cases stronger than traditional materials. They are also waterproof and non-inflammable, although at high temperatures they are liable to melt, and if the temperature continues to rise they may break down to release poisonous choking fumes. They are harmless, however, in ordinary use. You can cut and drill them without releasing anything harmful, and they will not injure wildlife. They are manufactured by the chemical industry; some of the substances that go into them are poisonous or otherwise harmful to health, and with the best will in the world all chemical factories leak a little. If you support efforts to reduce the industrial pollution of water and air to a minimum, there is nothing inconsistent about using materials that are harmless in themselves.

Roofing felt Roofing felt is a natural fibre which is felted and then impregnated with bitumen. Though it is inclined to tear, it is flexible and can be rolled. Lay it carefully to avoid tearing it, and secure it firmly with large-headed nails. Burning it releases dangerous fumes, but otherwise it will case no harm to you or the local wildlife.

Turf roofs Centuries ago people built houses from stone and timber and laid turf on the roof. In recent years the idea has been revived, at least experimentally, because of the difficulty of sealing flat roofs. The best and cheapest waterproof material is heavy duty plastic, but fitting it is difficult, since anything that pierces it weakens it and makes it liable to tear in strong winds, and when exposed to direct sunlight, ultraviolet radiation weakens it structurally. Turf may provide a solution to both problems. It is heavy, which secures the plastic sheeting and protects it from sunlight. It is also cheap, easy to obtain, provides good thermal insulation, is self-renewing, requires little mainten-

heavy-duty polythene
turf boarding

ance, and looks attractive. Your house may not have a flat roof, and even if it has you may feel uneasy about using such a seemingly unorthodox material, but why not try it on a garage or shed?

Choose a roof that is flat, or nearly so, and fairly large. Cover it with boarding to provide a smooth surface, then lay heavy-duty polythene to cover the entire roof, avoiding joins as far as possible — where there are joins, overlap the polythene and seal it with tape. Do not pin or nail the polythene, but secure it by laying turf in sections to cover the entire roof. You may need to water the grass in a very dry summer.

A turf roof will provide a habitat for many invertebrates, and cannot possibly cause pollution — an environmentally admirable solution.

Protection

Bitumen When crude oil is refined, its products are released one at a time starting with the most volatile. A mixture of methane, ethane, propane and butane appears first, as a gas. Then petrol (gasoline) appears, followed by paraffin (kerosene), diesel oil, and paraffin wax. After all that the residue is a black tarry substance called asphalt or bitumen. It melts when heated and sets hard when cooled. Its fumes smell delightful, but contain substances suspected of causing cancer, so try not to inhale them. While it is hard it will not harm wildlife, but in very hot weather it will soften and then it may become injurious to animals walking on it.

Being easy to apply and completely impermeable, bitumen has many uses. Mixed with gravel to add strength it makes a good surface for roads and paths. By itself it is used for roofing, but it has disadvantages. The stresses caused by expansion and contraction with heating and cooling will eventually make it crack, freezing and thawing of the water in the cracks will hasten the process, and the roof will become increasingly leaky. Bitumen can be patched to some extent, but once applied it can never be removed except by removing and renewing the entire roof. If the original roof was of slates, it may be cheaper to cover damaged slates with bitumen rather than replace them, but in the end it is less economical, so use bitumen with caution.

Creosote   There is a form of creosote used as medicine, but the creosote you use to preserve wood is made by distilling coal tar to yield an oily substance consisting of a complicated mixture of phenols and methylphenols. To be fully effective, the creosote must penetrate the wood thoroughly; painting the surface provides only partial protection. Wood decay is caused by fungi and insects — creosote kills both, and anything else exposed to it.

It is very poisonous, can cause burns if it splashes on to exposed skin, and is highly inflammable — in World War Two it was used as a fuel. Once wood has been treated, however, the creosote binds firmly to the wood fibres and does not leave them. Thus it causes no harm to wildlife other than the species against which it is designed to give protection, and can safely be used in the garden.

Fungicides in proofing materials Buildings can be damaged by weathering or by decay. Weathering is caused by wind, rain and frost, and most decay is caused by fungi. Protective outdoor paints are likely to contain fungicides, and all wood preservatives either are fungicides or contain them. Use them with care, observe the instructions on the container, and wash your hands, working clothes and brushes carefully when you finish. Do not dispose of small amounts left over at the end by pouring them on the ground or down the drain, and do not leave empty containers lying around where children, pets or passing wild animals may touch them. Seal containers firmly, leaving inside them any residues you may wish to throw away, and give them to the dustman. If you have more than one or two containers, ask your council for advice about disposal.

Weatherproofing The principle of weatherproofing is very simple: you protect one building by placing it inside another. In practice this means adding an additional outer skin to the structure, either all round or just on the most exposed side. Make sure the outer skin really is weatherproof, and that it does not trap moisture behind it to penetrate the inner wall. The materials used are generally harmless, but in some cases their production incurs an environmental cost. Weatherboarding is made from timber (*see* wood, page 63), and protective paints intended for application to walls are usually based on cement (*see* concrete, page 62).

Safety

Woodworking If you are a keen carpenter and spend your leisure time working with wood in an enclosed space, wear a face mask. You can buy one from any shop that sells industrial protective clothing. The inhalation of large amounts of wood dust may cause cancer of the nose.

Electricity generators

If you have vital electrical or electronic equipment that must run for long periods without even a brief interruption, you may need some kind of insurance against power cuts or failures. This may arise, for instance, if you live in the country and run long computer programmes. Freezers full of food, on the other hand, can survive most power cuts without harm, and in any case you can always insure the value of the food.

If they need any alternative to the mains at all, most people want a reliable reserve rather than an alternative electricity supply, and the only practical way to provide it is with a standby generator powered by a petrol or diesel engine. It will run more efficiently than a similar engine in a car because while it is running it works at a constant speed, but it must start reliably (and possibly automatically, under the control of a sensor that detects the first small drop in voltage that precedes total power loss).

Site the generator away from the house if you possibly can, ideally inside a well-built, well-ventilated outbuilding where it will be protected and its noise muffled. Generators are noisy, so remember the neighbours, but they are less polluting than vehicle engines.

Diesel engines Internal combustion engines (in which the fuel is burned inside cylinders) are limited in the fuels they can use. Diesel engines are the most versatile. Depending on its size and design, a diesel engine can run on gas or anything from paraffin to crude oil, provided the fuel is clean, free from water, sand and other impurities, and can be sprayed through a nozzle. Environmentally this gives it an advantage over the petrol engine. Refining petroleum uses energy and causes pollution around the refinery, so a less refined fuel is preferable to highly refined petroleum.

Unlike the petrol engine, the diesel engine has no ignition system, and this is the secret of its success. The movement of the piston compresses the air in the cylinder so hard that its temperature is raised above that needed to ignite the fuel. The fuel is then sprayed into the compressed air and each droplet ignites spontaneously.

The engine is very robust and reliable, but unless it is maintained properly it may inject too much fuel for the volume of air in the cylinder. When

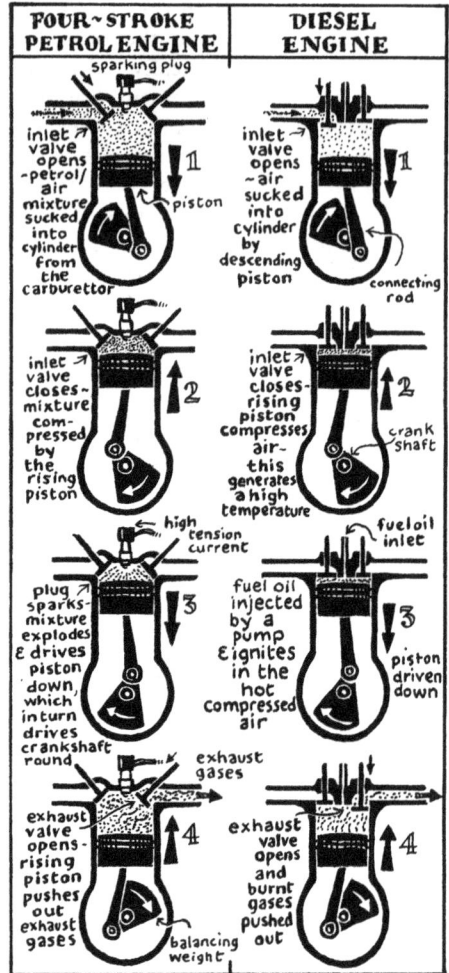

this happens the fuel is burned inefficiently, and some of it discharged as a thick black noxious smoke. Diesel engines in large vehicles are a major source of pollution, and these days diesel smoke accounts for most of the blackening of buildings in urban areas. The problem can be prevented by making sure the fuel nozzles are kept clean, clear of obstruction, and replaced when necessary, and that the fuel pumping mechanism supplying them is working properly.

A stationary engine used to generate electricity is easier to maintain than one installed in a vehicle.

External combustion engine An internal combustion engine is one in which the fuel is burned inside the cylinders. This limits the possible fuels to fluids that can be injected as a spray or vapour. An external combustion engine, such as a steam engine, burns its fuel outside the cylinders and consequently is able to use any source of energy, provided only that it can attain an adequate working temperature.

The steam engine is very inefficient, but not so the Stirling engine, invented in 1816 by Robert Stirling. For stationary engines such as electricity generators it has many advantages over the petrol or diesel engine. It can be adapted to run on any combustible fuel, electricity, or even solar energy if there is enough of it, and produces less vibration and is quieter than any internal combustion engine of similar size. What is more, it can achieve the highest thermodynamic efficiency possible for a heat engine, which means it converts more of its fuel into useful energy than any other type of engine.

Its working principle is simple. A closed system of piping carries a gas — these days usually hydrogen or helium — from a heat exchanger where the fuel is burned, to cylinders where the expansion of the gas drives pistons, and back again to the heat exchanger.

The environmental benefits are considerable. Being efficient, the engine is economical in its use of fuel, it is quiet, and because its fuel is burned separately from the engine, pollution is fairly simple to control.

Stirling engines are not easy to find, but if you need a generator it might pay you to ask around among the firms supplying generating equipment listed in *Yellow Pages*.

Transport

Consideration of engines leads naturally to thoughts of vehicles, and we all know that the family car is one of the more polluting of our inventions. What exactly do cars release into the air? Are there ways to make them cleaner? Are there alternative cars or alternatives to cars? Could we use cars less without making life impossibly difficult for ourselves?

Are your journeys really necessary? Road accidents kill about 5,500 people each year in Britain and injure many more. The manufacture and use of vehicles consume vast amounts of resources, and road transport is a major cause of air pollution.

The number of vehicles on the roads has been increasing for years, and it is tempting to assume that the trend will continue indefinitely and the problems will grow worse. This is almost certainly not what will happen. Most forecasters agree that in years to come we will travel less rather than more, because we will live differently. In Britain and America, and probably in other Western countries too, rural populations are growing while urban populations dwindle. When this trend is combined with new information technologies, a picture emerges of smaller cities but larger villages, with people working at or close to home and communicating electronically with their colleagues and employers, thus dramatically reducing the need to travel.

If people can work near to their homes, young people will not need to move away to find employment and families will be less dispersed. Friends and relations will be nearer, and journeys to visit them shorter. If people live in smaller communities spread more evenly across the country, journeys to the country and beauty spots will be shorter. If communities are more dispersed they will become more self-contained, so each community will have its own full range of shops, and shopping trips will be shorter. Electronic shopping using computers will reduce shopping trips still further.

CAR EXHAUST EMISSIONS (in descending order of amount)

① **Carbon dioxide** ~ The main product of the burning of carbon. In a confined space it can make the air seem stuffy. It is the most serious 'greenhouse' gas.

② **Water vapour** ~ A product of burning the hydrogen in a hydrocarbon fuel. It has no appreciable environmental effect.

③ **Nitrogen oxides (NO_x)** ~ Petrol contains some nitrogen which is oxidised when the fuel burns, but modern high compression engines operate at cylinder temperatures high enough to oxidise some atmospheric nitrogen, which joins the exhaust gases. In still, humid air and strong sunlight NO_x may combine with hydrocarbons and react to produce a range of compounds responsible for photochemical smog. NO_x may also be responsible for acid rain damage to trees.

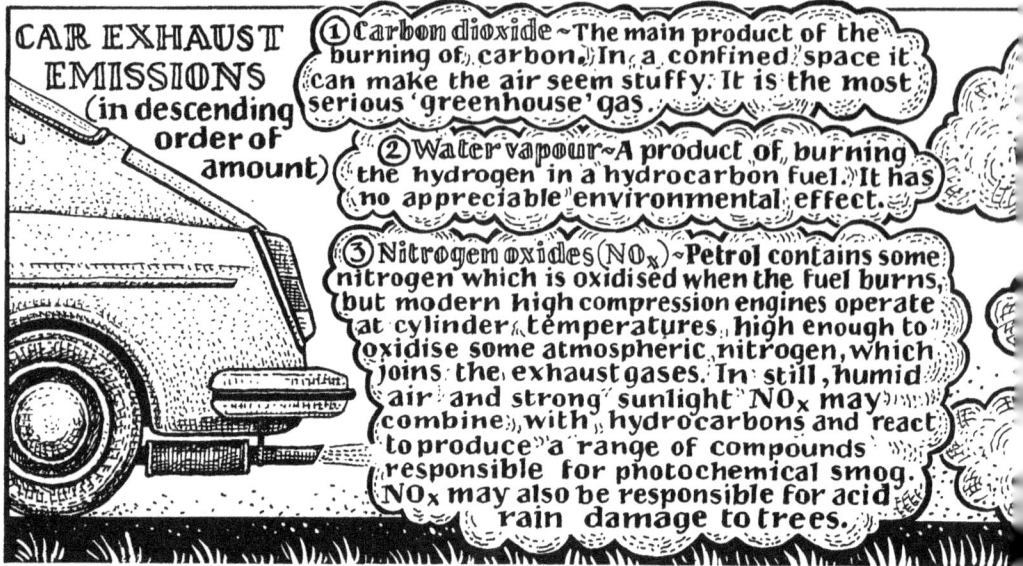

Pollution by cars

New cars or old? Without a car life in rural Britain is nasty and brutish. The present policy for public transport provides one bus every other Wednesday going in the wrong direction with fares that would make an oil tycoon check his bank statement. Country dwellers may be richly supplied with beautiful scenery, spiritually uplifting mud, and the tranquil song of peewit and chainsaw, but they tend to be short of negotiable currency.

Old cars cost less than new cars, so there are many old cars (and little yellow vans pensioned off by British Telecom) on country roads. Judging by the anti-nuclear stickers many of these ancient vehicles display, you might think an old car did less harm to the environment than a new one. This is the opposite of the truth. As it grows older an engine becomes less efficient, and the less efficient the engine the more fuel and oil it consumes and the more noxious the fumes it emits. In the last few years manufacturers have improved engine performance in order to economize on fuel and reduce pollution, and now they are modifying designs against the day when the addition of lead to petrol is forbidden.

You may not be able to afford a new car, of course, and must do the best you can, but as far as the environment is concerned new cars are better than old, and the older the car the dirtier and greedier it is likely to be.

Car exhaust emissions
(in descending order of amount)

Carbon dioxide The main product of the burning of carbon. In a confined space it can make the air seem stuffy. It is the most serious 'greenhouse' gas (*see* page 14).

Water vapour A product of burning the hydrogen in a hydrocarbon fuel. It has no appreciable environmental effect.

Nitrogen oxides (NO_x) Petrol contains some nitrogen, which is oxidized when the fuel burns, but modern high-compression engines operate at cylinder temperatures high enough to oxidize some atmospheric nitrogen, which joins the exhaust gases. In humid still air and strong sunlight

④ **Hydrocarbons** ~ Unburned hydrocarbons are emitted as exhaust smoke, consisting of a range of compounds many of which are known to cause cancer or are suspected of doing so. Hydrocarbons may also react with NO_x to cause photochemical smog and pollutants blamed for damaging trees.

⑤ **Carbon monoxide** ~ The partly oxidised product of burning carbon in a restricted air supply, carbon monoxide is emitted when engines are idling.

⑥ **Lead** ~ A very small amount of tetraethyl lead ($Pb(C_2H_5)_4$) is added to most petrol to increase its octane number and reduce engine knocking. The lead then joins the exhaust gases, some of it in a form that can be taken up by the human body and enter the bloodstream.

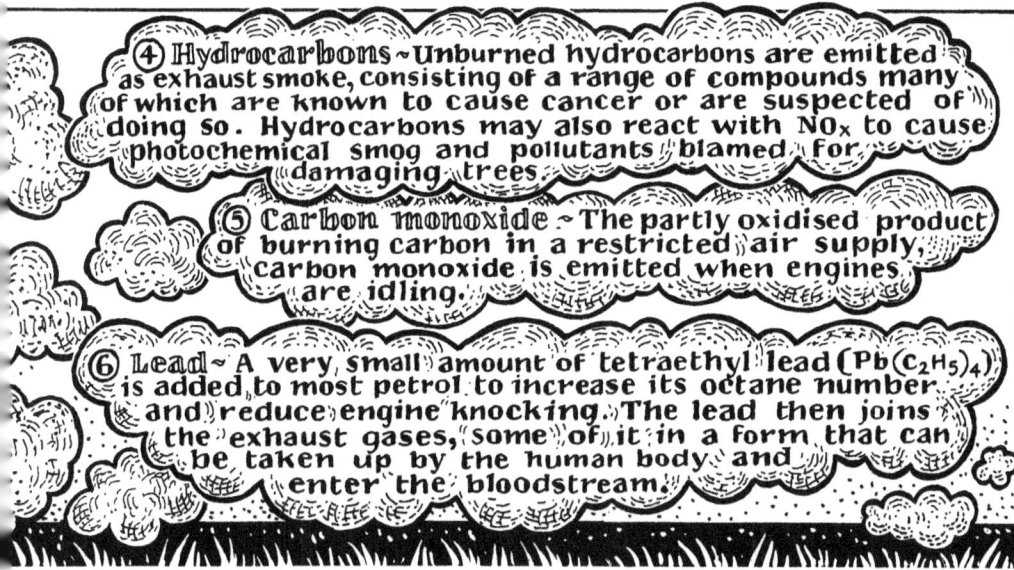

NO_x may combine with hydrocarbons and react to produce a range of compounds responsible for photochemical smog. NO_x may also be responsible for acid rain damage to trees.

Hydrocarbons Unburned hydrocarbons are emitted as exhaust smoke, consisting of a range of compounds many of which are known to cause cancer or are suspected of doing so. Hydrocarbons may also react with NO_x to cause photochemical smog and pollutants implicated in damaging trees.

Carbon monoxide The partly oxidized product of burning carbon in a restricted air supply, carbon monoxide is emitted when engines are idling (*see* below).

Lead A very small amount of tetrathyl lead ($Pb(C_2H_5)_4$) is added to most petrol to increase its octane number and reduce engine knocking. The lead then joins the exhaust gases, some of it in a form that can be taken up by the human body and enter the bloodstream (*see* page 74).

Carbon monoxide When you burn any fuel containing carbon, the carbon is oxidized. If it burns efficiently it will oxidize to carbon dioxide (CO_2), but if there is a shortage of air during combustion it will only partially oxidize to carbon monoxide (CO). Car engines tend to produce carbon monoxide when they are idling, which means that the gas sometimes accumulates in busy city streets.

Carbon monoxide is poisonous. If inhaled, it can pass from the lungs into the bloodstream, where it will combine with haemoglobin. Oxyhaemoglobin gives up its oxygen to body cells enabling them to respire, but carboxyhaemoglobin supplies no oxygen, thus inhibiting cell respiration. The victim will feel sleepy, and in extreme cases the poisoning can be fatal. This is why you should never run a car engine in a garage with the door closed. Without an adequate air supply, carbon monoxide emissions will increase and the gas will accumulate. To avoid producing needless carbon monoxide (and fuel), switch the engine off if you are in a traffic jam and know you will not be able to move for several minutes.

The problem, however, should not

be exaggerated. The concentration of carbon monoxide in the street is never enough to kill anyone. Mild intoxication reduces the oxygen supply to the brain, so the victim is not fully alert, but once the supply of fresh air is restored, recovery is rapid and complete, and there are no after-effects. Nor does carbon monoxide remain in the air for more than a few minutes before its oxidation is completed by ordinary atmospheric oxygen. Problems arise only in enclosed places such as streets surrounded by tall buildings, where the air is still and many car engines are idling in stationary or slow-moving traffic.

Lead in petrol Lead in large doses is poisonous, and its effects are well-known. Its effects in small doses are less well understood. We are all exposed to it, and have been for several generations, and no specific major effect has been noticed. It is possible, however, that young children can be harmed by quite small amounts, some of which they receive by inhaling air contaminated with car exhausts containing lead (*see* car exhaust emissions, page 72). The evidence of harm is uncertain, but under pressure from campaigners governments have agreed that it would be prudent to reduce the amount of lead in petrol, and eventually to prohibit leaded petrol altogether.

In all EEC countries, new cars will have to run on unleaded petrol from 1990, and some unleaded petrol is on sale now. Do not use it, however, unless you are certain that your car engine will accept it. Lead is added to increase the octane number of the fuel, which prevents knocking (the explosive combustion of some of the fuel in the cylinder, causing a unpleasant noise, damage to the spark plugs, and overheating). Engines can be designed to use unleaded petrol, but it can damage those designed for high octane fuel.

Alternatives to petrol

Petrol burns readily with a hot flame, and at ordinary temperatures it is a liquid, which makes it easy to transport. It is an ideal fuel, its only disadvantage being the gases which are the products of its combustion (*see* car exhaust emissions).

When oil prices rose in the 1970s, research began in earnest to find a cheaper alternative. Since then, however, prices have fallen again, so the search has lost some of its urgency. During those years various possibilities were considered, and work on them has continued, though with less publicity. When oil prices rise again, as they surely will, one or another of them will emerge as the winner. The most probable among them are considered below, but first we must never forget the most environmentally benign of all vehicles, the bicycle.

Bicycles The bicycle is said to be the most efficient means of transport ever invented. The claim is based on the distance the vehicle and its rider can travel on a given amount of fuel — in this case a share of the cyclist's lunch. On this basis the bicycle beats all challengers, and modern bicycles are even better than their ancestors. The best of them have frames made from lightweight tubes, elliptical in cross-section to improve aerodynamic efficiency, and equipped with up to twelve gears where a generation ago three gears were a luxury. Although limited in the amount of luggage it can carry, and too slow to be practicable for long journeys, provided the cyclist is strong it is nevertheless the best means for moving around a local area. Those who cycle in urban areas also need to be brave, assertive, and not too bothered by traffic fumes.

Electric cars Electrically-powered

cars are almost silent, cause no air pollution, and in Britain are exempt from road tax and do not have to pass MOT tests. Unless they receive their power from outside, however, they must be driven by batteries, which are very heavy. They work well enough in milk floats which travel short distances, slowly and with frequent stops, but they cannot travel far, fast, or up steep hills.

Batteries are now being designed that are lighter, more powerful, and easier to recharge. One using aluminium, for example, might drive a family-sized car 250 miles before it needed water to replenish its electrolyte, and 1,250 miles before its batteries needed new sheets of aluminium. It works out at just over six miles per pound of aluminium, and the batteries would need no other recharging — by the mid-1990s your new car may be running on aluminium!

Flywheels Flywheels look promising, though flywheel-powered cars have never yet been built. The theory is that a battery-powered electric car running at constant speed spins a heavy flywheel, and the flywheel supplies power to the wheels. The car would coast down hills, allowing the flywheel to drive a generator, partly recharging the battery.

A hybrid car is also possible, with a small petrol or diesel engine running at constant speed to keep the batteries charged.

Gas as a fuel During World War Two some vehicles were modified to run on town gas (mainly carbon monoxide). The fuel was stored in a large bag on the roof. More recently, bottled gas has been used — this is LPG (Liquefied Petroleum Gas), a by-product of oil refining. Its principal advantage in Britain is that it is not taxed like petrol, so it is cheap. The main disadvantages are the cost of modifying the engine to

burn it, and the weight of the gas bottles.

One chicken farmer used to feed poultry manure into a fermenting vessel to produce methane, and ran his car on that. It might work, but you will need a poultry farm, since nothing smaller will provide enough manure. If you do decide to try it, be careful. In the 1970s many people tried to produce methane from manure. Almost invariably the 'digester' either failed to work at all, or did work and exploded (*see* methane digesters, page 78).

Hydrogen as a fuel Dramatic pictures of blazing airships have convinced many people that hydrogen is dangerous; in fact it is very safe. Being lighter than air, it rises rapidly when it escapes, and in a hydrogen fire the flames leap upwards, away from the people. As a gas it cannot be splashed or spilled to spread the fire, and it burns at a lower temperature than petroleum-based fuels, so the damage and injuries it causes are less severe. When it burns the only combustion product is water vapour, making it the cleanest of all fuels.

Unfortunately it also has real disadvantages. It is obtained from water, using energy to break the bond to separate water (H_2O) into hydrogen (H) and hydroxyl (OH), two H atoms then combining to form a hydrogen molecule (H_2). But it takes nearly as much energy to obtain the hydrogen as is released when the hydrogen burns (recombining with oxygen to make water, $2H_2 + O_2 \rightarrow 2H_2O$), so when the energy involved in getting the fuel to the consumer is taken into account, the operation is not very economical.

As if that were not bad enough, the engine needs a lot of hydrogen since it burns at a relatively low temperature, and being a light gas it is not easy to store. It can be held under pressure, but then the fuel tank must be a pressure vessel, and therefore heavy. It can be made to form a temporary association with various minerals, so the fuel is like a 'gravel' that gives up hydrogen when warmed — then you have to have a 'fuel tank' big enough to carry the 'gravel'. Hydrogen might be used in a transport aircraft with a huge fuel tank containing fuel that is lighter than air, but although such aircraft have been designed, no one seems particularly eager to equip the airlines with hydrogen tanks.

The garden

A source of energy

If big power stations worry you, should you think about generating your own electricity from wind, flowing water, or even sewage? What are the advantages and disadvantages?

Wind generators It is perfectly possible to generate electricity on a small scale for household use from wind power, but it is not cheap. There is not much point considering it unless you live in a very remote and very windy place where no mains electricity can be persuaded to reach you.

The equipment is far removed from the traditional windmill. It consists of a metal tower, two or three rotor blades, gearing and governing devices, batteries for storage, and cable. The tower must stand on open ground 4.5-6 metres above the height of trees and buildings. Every wind generator has a 'rated capacity', which is the electrical output of which it is capable. It will achieve

this only when the wind blows at the speed for which the rotors are designed. In practice you may expect to generate about 700kWh per year for each kilowatt of rated capacity, so on average a 4kW rated system will provide about 8kWh a day.

Go round your house, make a note of every electric light and appliance

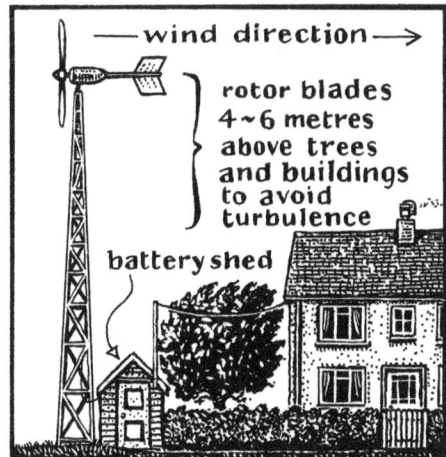

— wind direction →

rotor blades
4~6 metres
above trees
and buildings
to avoid
turbulence

battery shed

Labels in illustration: overshot water-wheel / 4m diameter turning at 8 rpm / 35 cm wheel turning at 500 rpm / pelton wheel / a jet of water drives a wheel of small buckets round at great speed / generator turning at 1800 rpm / 30 cm turbine turning at 300 rpm / cross-flow turbine

with its power requirement (one bar of an electric fire uses one kilowatt, for example), multiply each one by the number of hours a day it is in use, add up the answers, and you will know how much electricity you need each day in kWh (kilowatt hours). Multiply this by 365 (the number of days in the year), divide the answer by 700, and the final number will be the rated capacity of the system you will need.

Hydroelectric power If there is a reliable and fast-flowing river running across your land you may be able to take power from it. The principle is that of an old-fashioned watermill, but the equipment is a scaled-down version of that used in a hydroelectric power station. Water will be diverted from the river, made to fall across a set of turbines, and returned to the river at a lower level. The spinning turbines will be connected to a generator.

The installation will not be cheap, and you will have to obtain permission from the water authority, but the equipment will require little further maintenance and the output will be reliable. The system must be designed as a whole and tailored to the power available from your river, so do not be tempted by offers of second-hand equipment.

Methane generators Natural gas consists mainly of methane, so a device that

produces methane which you can hook up to your cooker has obvious attractions. This is what a methane generator is supposed to do, and many biogas plants have been installed in the tropics and sub-tropics, especially in India; many British sewage treatment works have them too. A few years ago small ones were also in vogue in Britain. At least, there was much talk about them.

A methane generator or 'digester' uses human sewage or livestock waste as its raw material. It yields combustible gas which is rich in methane, and a slurry which can be used as a fertilizer. It is more difficult than it sounds to achieve the fairly precise biochemical conditions necessary for the digester to work, but a more serious trouble is that it needs a very large amount of raw material if it is to produce enough gas to be of any use. Some people have used the entire output from a poultry farm, and in India each plant is designed to serve and take the sewage from a village of hundreds of people. It is simply not practicable for an ordinary family, and even if it were, in Britain there is a second difficulty. The raw material is digested by bacteria, which must be kept warm. They work best at 30-40°C, and cease working if the temperature falls much below 10°C. This means that for much of the year the digester has to be heated, which consumes energy and negates the energy produced by the system.

Food production

Some people enjoy growing their own food; others find a vegetable patch ugly and gardening tedious. How can you improve the appearance of the cabbage patch and cut down on the work?

Is the soil poisoned? In urban areas, and close to main roads, the air is polluted, and some of the pollutants — including lead — fall out on to adjacent land. The ground on which houses are built may formerly have been used as the site for a factory, and industrial waste may have been tipped on it. Before you start growing food crops in your garden it may be as well to find out whether any pollutants in your soil will be taken up by plants. Your local authority horticultural adviser should be able to tell you how to set about getting soil analysed — you will find the address and telephone number in the *Telephone Book*.

Planning a layout Most gardens are not planned at all. They sort of happen as carefree owners pop a bush in here, lay a bit of lawn there, and make it up as they go along. If you inherit an existing garden, making radical changes to its design may involve a great deal of work. All the same, where it is possible planning will save effort and money.

The first step is to decide what purpose the garden is to serve. Is it to be mainly decorative, a joy to behold and maybe open to the public, with an admission charge? Is it to be a place where you entertain, and (may you be forgiven) hold barbecues? Is it to be a place where you can pursue your outdoor interests other than gardening? Is it to produce food? Is it to become a refuge for wildlife? Or is it to be some mixture of several of these? The answer will depend to some extent on the size and shape of the garden, but even with only a tiny area a little imagination can go a long way.

Vegetables in a small space Growing vegetables is hard work, although your own produce tastes better for the knowledge that you grew it yourself. Apart from the work, the main disadvantages are economic and aesthetic. There are correct times and conditions for sowing and planting out, and therefore each crop has its season, which is the same for the amateur as for the professional. Your vegetables will be ready to eat at the same time as those in the farm and market gardens, so you will be harvesting just when shop prices are at their lowest. Runner beans, with their intensely green leaves and brilliant red flowers, are indeed beautiful, but only the eye of true love can dwell

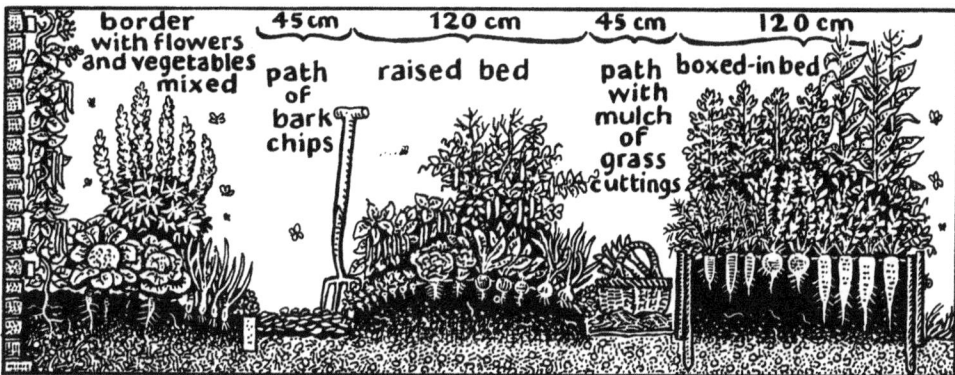

border with flowers and vegetables mixed | 45 cm | path of bark chips | 120 cm | raised bed | 45 cm | path with mulch of grass cuttings | 120 cm | boxed-in bed

enraptured on a row of partly-cut, leaf-dropping, dog-eared brussels sprouts. Some vegetable crops are attractive, but many are unredeemedly drab.

If you can cram the vegetables into a much smaller space you will solve several problems at once. They will be easier to reach and manage, and less obtrusive. You can even make them more decorative. Whatever you do, intersperse the vegetables both one kind with another and with decorative plants to avoid ugly rows of brassicas. This will bring an added advantage: the better the ground cover and mixing of plants, the less space there is for weeds and the fewer the opportunities for pests and diseases. While mixing crops in this way has disadvantages for the large-scale grower, it is more efficient on a small scale, reducing the amount of work needed and improving the quality and yield of the crop.

Try growing vegetables in narrow side beds against a wall or fence, the tallest at the back and the shortest at the front, just as you would shrubs or flowers. Further into the garden try raised beds contained by small decorative fences, using earth or compost to build up the ground level to about knee height — the added depth of fertile growing material will produce splendid crops.

Greenhouses The simplest and cheapest greenhouse consists of a rigid frame covered with clear heavy-duty polythene. The polythene skin is held in place with screws or bolts with very large padded washers, and adhesive tape. You can buy such greenhouses in kit form, or you could improvize one. It can be of the traditional shape with walls and a ridged roof, or have a frame of semi-circular hoops to make a tunnel. The plastic has to be repaired more often than glass and eventually you have to renew it, but it is just as effective as glass.

Buy only those plastics intended for horticultural use, including plastic pots and plastic-covered supports and wire, and keep a plastic greenhouse well-ventilated at all times. There is some evidence that plants may be harmed by volatile chemicals leaking from some plastics, and plants have even been damaged by harmless covers supported on plastic-coated wire. The harmful chemicals are phthalates, used to make cellulose acetate, PVC, and other plastics more flexible; other chemicals may also be involved, since plants have been damaged by plastics containing no phthalates. Not all plants are affected, but tomatoes, cucumbers, peppers, aubergines and brassicas are. The doses that injure plants are far too small to harm humans.

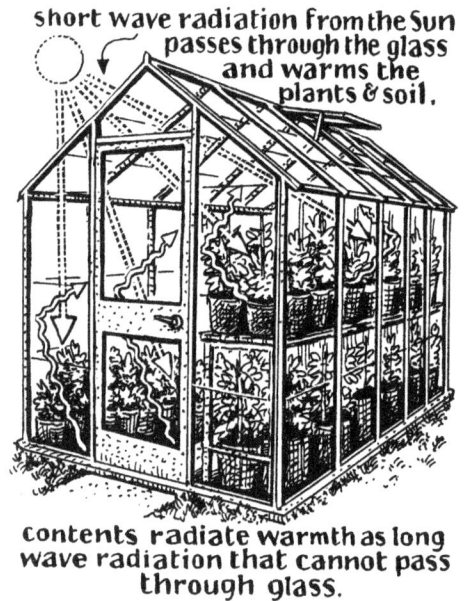

short wave radiation from the Sun passes through the glass and warms the plants & soil.

contents radiate warmth as long wave radiation that cannot pass through glass.

Greenhouse heating and ventilation A greenhouse is a warm place for two reasons: shelter from the wind, and the 'greenhouse effect'. Air inside it is trapped, so as the sunshine raises the temperature warm air cannot be replaced by cooler air, but neither can it escape easily. Short-wave radiation from the Sun passes through the glass or plastic of the walls and roof and warms the contents of the greenhouse, which start radiating their warmth, but as long-wave radation which cannot pass through glass or plastic.

Because heat can enter the greenhouse much more easily than it can leave, on a hot day the temperature in a greenhouse can rise very high. At more than about 40°C respiration is inhibited and, starting with the most sensitive species, plants die. You must prevent your greenhouse from overheating, and if you are designing your own, remember to include a window that will open, preferably in the roof to allow hot air to rise and escape. More elaborate greenhouses have opaque screens like curtains that roll down part of the way across the roof, operated by an electric motor linked to a thermometer so they work automatically. The screens may also cover the roof at night, to reduce the small loss of heat by conduction caused by the warm air inside warming the windows themselves.

Plants will not grow at temperatures less than about 5°C, and many are sensitive to being chilled, so in winter you may need to heat your greenhouse. There are electric heaters, some of which run heating elements through the soil and heat it directly. Some commercial greenhouses are connected to a hot-water central heating system with a boiler, but the simplest device is a paraffin heater. As the fuel burns, carbon dioxide is released as a combustion product, which is good for your plants. They need carbon dioxide for photosynthesis, and ordinary air contains rather less of it than they can use. Give them more and they grow faster and larger — some commercial growers pump carbon dioxide into their greenhouses for this reason.

Provided it has an earth floor, or you are just building it and can play around with a concrete floor before it sets, you can economize on heating bills in the

greenhouse by recycling old glass bottles. Bury the bottles in the floor so their tops are just below what will be the surface. Fill them all about three-quarters full with water (to allow room for expansion should the water freeze), seal them, and finish the floor. They will not break because you will be walking across their tops, and it is almost impossible to break a bottle that way. The water will absorb heat while the inside of the greenhouse is warm, and release it when it is cool.

Nutrient film technique A fairly simple way to produce large yields of suitable greenhouse crops very economically, nutrient film technique uses no soil at all. If you are handy with tools you can improvise a system for yourself. Use alkathene guttering, flat-bottomed if possible, to make a series of troughs linked to more guttering across the open ends. Arrange this framework of gutters on supports so they are all on a gentle incline, with the highest corner diagonally opposite the lowest. Place a tank so that it can feed water to the highest corner, and make holes in the top gutter so the water will trickle into each of the descending gutters, covering them with a film of moisture, and flowing out into the lower gutter. Fit an electric pump and piping to pump the water back up to the tank at the top. Add plant nutrients to the water (you can buy suitable

circulation of nutrient

ready-prepared solutions), place your seedlings side by side along the trough, close together so they can support each other, and they will be fed and watered from below. Once growing, the plants need no further attention unless they are attacked by pests or disease. Obviously there can be no weeds. The technique is most suitable for low-growing leafy plants like lettuces, but you can experiment with others.

Herbs A small and well-planned herb garden is attractive, needs little attention apart from weeding and harvesting, and over the years you can build up a large collection in a very small space. Some can be used directly in the kitchen, but your garden will almost

certainly produce more than you can use fresh, so you will be able to dry some for long-term storage. Apart from culinary herbs there are others with medicinal or cosmetic uses and you might try growing some dye plants as well.

Most herbs are aromatic and produce coloured flowers, and this is the other reason for growing them — they are attractive to nectar-feeding insects. Once the herbs are established your garden will be filled with bees, hover-flies, butterflies, moths and many smaller insects, together with the predators that hunt them.

Fertilizers Plants are made from substances they obtain from the air, water and soil. When they die those ingredients are returned and can be used by other plants, but if you remove a plant and eat it that cycle is broken. If you want to grow crops you must replace the nutrients taken from the soil with fertilizers — either factory-made ones or organic matter which you buy or produce in your own compost heap. It sounds straightforward, but it is not quite so simple. Plants take up nutrients only after they have been dissolved in water, so quick-acting fertilizers are readily soluble. This means that they are liable to be washed out of the soil. They do not then benefit the crop but can pollute water, so be careful how you use them.

That is not all. Fertilizers favour a limited number of species, which can

Harvest herbs on a warm and sunny summer morning when the dew is dry... but before the leaf oil content is reduced by the sun's heat. Hang up to dry out in a warm place like an airing cupboard...or more quickly on trays in a warm oven — this takes only a few minutes. airtight dark glass container. Crumble the dry leaves and store ready for use.

take up nutrients and grow rapidly, crowding out their competitors. If you apply no fertilizers your crops will suffer, but the slower-growing plants will survive better, leading to a much greater diversity of species, and where there is a greater diversity of plants there will also be a greater diversity of animals. Use fertilizers to grow crops, but avoid them in areas where you want to encourage wildlife, where they will do much more harm than good.

Composting Organic wastes should whenever possible be returned to the garden, either by burying them under the soil or by composting them. Wastes from the garden itself, kitchen scraps, eggshells, plate scrapings and any other soft material is suitable. Woody material will rot very slowly, bones will eventually disappear, dead leaves can be disposed of this way but contain little plant nutrient, and some leaves — such as privet — rot very slowly.

You can recycle wastes by burying them in trenches dug into the flower or vegetable beds beneath a good layer or topsoil, or you can compost them. To do this, make two containers the same

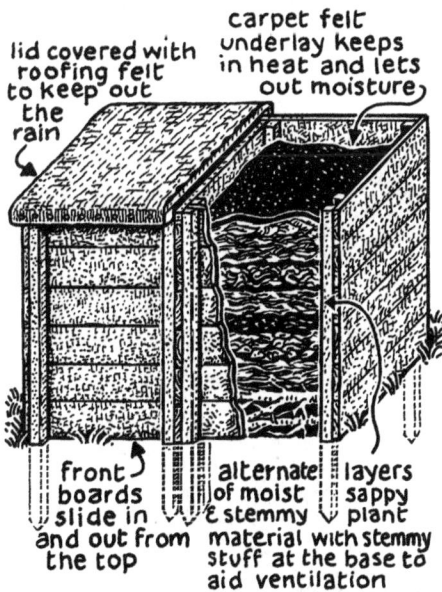

lid covered with roofing felt to keep out the rain

carpet felt underlay keeps in heat and lets out moisture

front boards slide in and out from the top

alternate layers of moist & stemmy plant material with stemmy stuff at the base to aid ventilation

sappy

size and side by side, with fronts you can remove for easy access. Place branches on the bottom to provide air spaces. Fill the first container with a sequence of layers of about 15 centimetres of plant material, a dusting of lime (which you can buy cheaply from any garden shop or centre), and a layer ideally of animal manure, but failing that a dusting of nitrogen-based fertilizer or a proprietory 'compost starter', any of which will supply nitrogen or nitrogen-fixing bacteria.

When the container is full, cover it and wait. Small animals and fungi will feed on it, breaking it down into smaller fragments, and before long the heap should have a large population of earthworms. Bacteria will take over at the end of the operation, digesting organic compounds and in so doing changing them into the much simpler compounds which can be taken up again by plant roots. The chemistry of a compost heap is extremely complicated and not fully understood, but we do know that bacterial decomposition involves some reactions that generate heat. The amount of heat produced by each bacterial cell is minute, of course, but the heap contains countless billions of such cells, enough to heat the entire heap — you may even see steam rising from it. The warmth may germinate many weed seeds, but the weed will be killed later.

The bacterial activity is self-limiting. After a few weeks, the precise time depending on the size of the heap, the bacteria will have consumed most of the nutrients available to them and their numbers will start to fall. As this happens the heap will cool.

When it has cooled, empty the heap into the adjacent container so what was on top is now at the bottom. This mixes the heap and smothers and kills most of the weed seedlings. The disturbance and aeration will encourage bacterial activity to start again, so the heap may heat for a second time, though less strongly. After a few more weeks the entire heap should have turned into a sweet-smelling, brown, more or less weed-free and friable compost, ready for immediate use.

Irrigation Never water plants in warm sunshine — the water on their leaves will act as lenses and cause burning. As far as possible water the soil rather than the plants, and drench the soil thoroughly, since water will drain quickly through dry soil, most of it disappearing before the plants have time to benefit. If you have a large area to water, discuss your irrigation problems with the water authority. You may be advised to build a tank large enough to hold a one-day supply, and fill it from the mains when demand is low — you may be charged for the water. In Britain you must have a license to take water for irrigation (though not for domestic use) from a borehole, stream, river, pond or lake, even if the water lies entirely on your own land.

Water butts You need no license, and will not be charged, for the rain that falls on your garden, or for that matter on the roof of your house. No one will complain if you can distribute the rain a little more evenly. All you need is a large water butt, and the larger the better. Pick the largest roof area on your house, find the pipe that carries the rain from the roof gutter down to the drain and disconnect it at an appropriate height, place your water butt beneath it, and finally fix an overflow pipe to carry the surplus from the top of the butt back to the drain. You will be surprised how fast the butt fills, even during the shortest shower.

Keep the top of the butt covered or it will provide a perfect breeding site for mosquitoes and other insects which spend their larval stages near the surface of water (it will be too deep to attract dragonflies or damselflies).

After a time the water will acquire a large population of bacteria and will smell sulphurous when you draw it. Leave the water to stand for a few minutes and the smell will disappear.

The water is excellent for irrigating the garden, but on no account should you drink it, because it will make you very sick — or worse. If you are something of a mechanical wizard you should be able to pump water from the butt through a hose, filtering it first so it does not clog the pump. If you are not you will have to carry a watering can like the rest of us.

Bonfires Most of us love a garden bonfire, smouldering the dead leaves into aromatic clouds that perfume the autumn air. Our enthusiasm may wane a little at some of the more pungent smells the fire can release when it is fed household refuse, and it may fade still more when we learn that even the delightful odour is composed of substances most of which can cause cancer. Put very simply: bonfires are bad news, but sometimes necessary. Compost as much refuse as you can, and you can add dead leaves to the compost heap, give the dustman as much as he will take, and when a bonfire is necessary, minimize the nuisance it causes.

Use an incinerator to raise the base of the fire above the ground, provide ample ventilation, and prevent the fire from spreading. Use whatever dry and highly combustible rubbish you can find to give the fire a hot, fierce centre so it burns quickly. This will reduce the time it takes to get rid of the rubbish, and the hotter the fire the more completely and cleanly it will burn. Try to avoid letting it smoulder, and try not to inhale the smoke.

Weeds and pests

They used to say that a weed was simply a plant in the wrong place, and a pest was an animal in the wrong place or misbehaving. It is not that simple. As the following sections show, a plant is not really a 'weed' nor an animal a 'pest' until it shows signs of causing serious damage. Most wild invaders of the garden are harmless, but attempts to exterminate them may not be.

If you feel obliged to use pesticides, it is as well to know about the armoury of weapons available. Use them carefully to avoid accidents, and before you use them at all pause to consider whether there may not be a safer alternative.

Weeds When you disturb ground by clearing away the vegetation and exposing the bare soil, then loosening the soil by digging, forking, or even pulling plants out by their roots, certain plants benefit. Their seeds, waiting in the soil for just such an opportunity, will germinate and grow rapidly. Bits of broken root will produce new growth, and root-spreading plants will invade from nearby. These opportunist plants are 'weeds', and if you want to distinguish between a 'weed' and any other kind of wild plant, look at what appears first on a demolition site or on piles of soil heaped beside a construction site.

Many of the opportunists are shal-low-rooted annuals. They are easy to remove and are killed when the soil is turned over and they are buried. Others are more persistent and root more deeply, like docks, or spread like nettles and couch grass.

All unwanted opportunists are best dealt with by removing or attacking them physically. Farmers may need to use herbicides, but there is no need for them in a garden. You do not need to remove all the 'weeds': they provide food for many insects, and some of the insects will take to eating your crops if you deny them any other food. 'Weeds' also help to keep the soil moist and shaded in hot dry weather, and aerated and sheltered in wet weather. The fashion for clearing every last 'weed' to leave ground 'clean' and neat is now considered extreme and at best pointless. Too many 'weeds' are bad news, but bare ground is worse.

Pests A 'pest' is an animal that devours so much of your crop that you face starvation or economic ruin. If you treat occasional passers-by as mortal foes and launch a full-scale chemical war against them you may do considerable harm while achieving nothing.

The first and most important task in dealing with pests, therefore, is to decide whether they are pests at all, and there are several gradations. It is necessary to be able to identify the commoner animal species (and if the animal is not common, then by definition it cannot be a pest). Slugs, for example, are widely regarded as pests, but in fact only certain species feed on food crops, and Britain has many species. There is no point at all in killing harmless individuals.

Having decided the animal belongs to a 'pest species' you need to judge whether it is present in numbers large enough to constitute an infestation. This is a little more complicated than it might seem, since the few individuals you find may be about to breed and reproduce prolifically. It will help greatly if you learn something about the way they live, and what you learn will also help you to take appropriate action if there really is a problem.

Most insects are vulnerable to attack only at certain times in their lives and they all have non-human enemies, so random assaults — especially with pesticides — may actually encourage them.

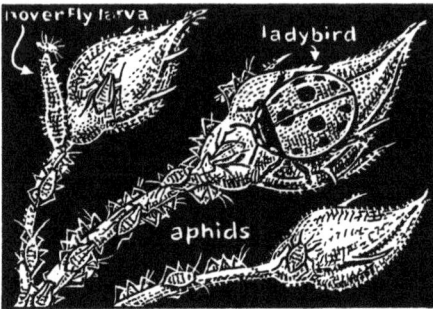

Aphids Greenfly, blackfly and plant lice are tiny insects. They mate late in the year, the females lay eggs on suitable plants, and then they die. The eggs hatch when the long warm days return. The young are all female, all wingless, and are called 'stem mothers' because they start reproducing at once.

Their young are produced parthenogenetically (without involving a male), thus are all female, and are born alive and active at the rate of a litter a day. They too reproduce, and when a plant becomes crowded aphids with wings are born and these migrate to another plant. In late summer they start to produce more winged forms, and some males. Mating occurs, eggs are laid, and the cycle begins again.

Spring is the time to look for them. If you see just a few on a plant they will be 'stem mothers', and if you are quick you can kill them before they start reproducing, even if it means removing and destroying the affected plant. When plants are covered in aphids it is really too late, though you can try

using an aphicide on them (*see* insecticides, page 91, for chemical remedies).

Pea and codlin moths The pale wriggling grubs you sometimes find when shelling peas are the larvae of the pea moth (*Cydia nigricana*). They are close relatives of *Cydia pomonella*, the codlin moth, which prefers apples, and both are more distantly related to many leaf-rolling moths, which can defoliate entire trees.

Pea moth larvae spend the winter in the soil safely wrapped in cocoons, and pupate early in May to emerge as adults between late May and July. The adults do not feed, so it is difficult to poison them. They mate and lay eggs, which hatch 9-16 days later. The young larvae then seek pea pods, which they enter to feed on the peas, remaining inside the pod for about three weeks. After that they work their way out, drop to the ground, bury themselves in the soil, spin their cocoons and wait for the following spring.

They are sheltered and secure from all insecticides for the whole of their lives except for the short period, often less than one day, between the time when the eggs hatch and the larvae disappear inside the pea pods. The more enlightened professional growers use traps with sticky floors, baited with the chemical by which females attract males for mating, to catch males. When males start arriving in large numbers

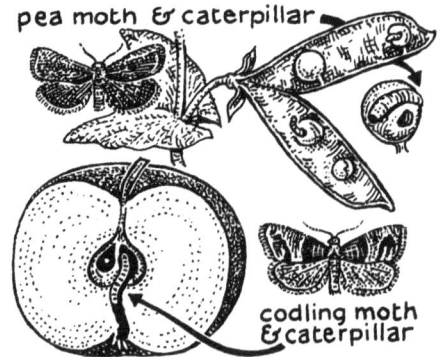

they wait about a week, spray to catch the emerging larvae, and spray again three weeks later to catch larvae leaving the pods, each time using a short-acting insecticide.

The old-fashioned alternative is to drench all the pea plants in a long-acting insecticide and hope to kill larvae as they make their way to the pods. This is undesirable because peas rely on insects to pollinate their flowers, so spraying can reduce the crop.

The modern technique does not work with the codlin moth, which lays its eggs inside the fruit, but scientists have identified, analyzed and synthesized a chemical substance produced by the female when she lays eggs. It attracts predators that feed on the eggs and larvae, and it deters females from laying eggs at sites that have been used already, but it poisons nothing.

Pesticide sprayers The conventional pesticide sprayer is essentially a pump with a nozzle; its size may vary, but not its operating principle. The active pesticide is dissolved in a liquid, which may be water or oil depending on the pesticide, poured into a reservoir, pumped from the reservoir through a pipe, from which it is sprayed through a narrow nozzle.

The spray itself consists of drops of liquid of many different sizes, and the larger of them usually fall to the ground as a continuous dribble from the nozzle. The sprayer is responsible for much environmental damage, and is very inefficient. Some plants, and some parts of each plant, are dosed much more heavily than others. Much of the spray misses the plants altogether, and of the pesticide that finds its target a great deal drips from the plants on to the ground. The undersides of the leaf are missed altogether. The pesticide on the ground may kill whatever lives there, including the emerging crop and the invertebrate predators that spend the daytime in the soil and climb the

Correct use of insecticides

If you must use an insecticide, there are some important rules to follow:

• Make sure you really have an infestation, not just a few visitors.

• Identify the pest.

• Buy a product designed specifically for that pest. As far as possible avoid all-purpose insecticides.

• Read the instructions on the label and follow them to the letter.

• If you have to dilute the product do not make a more concentrated solution than is recommended; it is unlikely to be more effective and may be dangerous.

• Do not spray more often than the instructions tell you.

• Keep unused insecticide in its original containers. Children could mistake it for something they can drink, and many have.

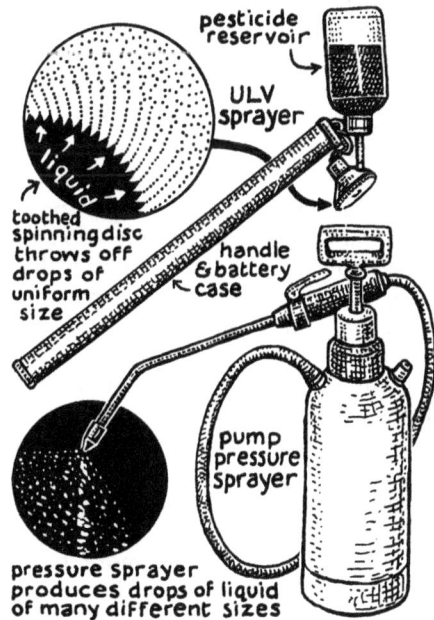

pesticide reservoir

ULV sprayer

toothed spinning disc throws off drops of uniform size

handle & battery case

pump pressure sprayer

pressure sprayer produces drops of liquid of many different sizes

plants at night in search of prey. Over-dosing of leaves can block the stomata, injuring the plant. Insect pests will be killed if the spray hits them, but those sheltered from the drops will escape. Others will receive a sub-lethal dose, and if they have thick exoskeletons or the inherited ability to survive the poison, the elimination of their preda-tors will allow them to form the nucleus of a population resistant to that particular pesticide, and possibly others like it.

It is possible to spray more efficient-ly if you can find a better sprayer. Ultra-low volume (ULV) sprayers use a torch battery to power a small electric motor that spins a toothed disc, rather like a cogwheel. A highly concentrated solution of the pesticide (liquid foliar feeds can be applied in the same way) is fed to the spinning disc, thrown to the outer edge, and leaves from the teeth as filaments that break up immediately into droplets which are all the same size. The pesticide moves through the crop as a mist, coating plants on the upper and lower surfaces, evenly but thinly. A ULV sprayer can achieve better results than a conventional sprayer with one tenth or less of the amount of pesticide.

The Electrodyn sprayer, developed by ICI, is even more efficient. It uses between 0.25 and 0.5 per cent of the amount of pesticide an ordinary spray-er uses. It achieves this by imparting an electrical charge to the droplets so they adhere to the plant surfaces rather than dripping from them. The Electrodyn is made only in a farm size at present, though if ICI is pressed hard enough it may produce one small enough for gar-deners. Why not write to them?

Obviously if you use less pesticide there is less opportunity for it to harm beneficial or innocuous plants and animals, so sprayers such as these are greatly preferable to the more tradi-tional types. They may not be easy to buy, but do ask at your garden shop,

and if that gets you nowhere, ask your county horticultural advisor, whose address and phone number will be in the Telephone Book.

Alternatives to chemicals Farmers who depend on their crops for a living may need to spray them (though many people would dispute the matter), but any licence they may have to use insect-icides does not necessarily apply to amateur gardeners. If you do not face starvation or ruin, are the pests in your garden anything more than an irritating inconvenience? Should you kill animals just because they irritate you?

Obviously you should be prepared. Cover fruit trees and bushes with nets, for example, to keep out birds. Beyond that, the first and most obvious alter-native to spraying is to do nothing at all. If the inactivity bothers you, try a physical assault. Brush the offending pests away from the plant, knock them off with water or, if there are only a few of them, pick them off by hand. If only one or two annual plants are affected, sacrifice the plants by removing and destroying them.

On the other hand, you might reflect that while your plants may be food for pests, the pests are food for other animals, some of which are interesting and attractive. Bats devour insects by the billion; ladybirds and lacewings are voracious carnivores of insects smaller than themselves. Centipedes in the soil and spiders in and out of it eat insects, and the larger hunters are food for still larger animals. Hedgehogs feed on invertebrates of all kinds, and have a special liking for snails and slugs, and so do many garden birds. If you feel you have to do something, look around for small predators such as ladybirds (but not hedgehogs, bats or other mam-mals) and move them to the infested plants. They will love you for it. This is a kind of 'biological control', and sometimes it works.

If at last you decide you must use an

insecticide, choose one that promises to poison the pest but not everything else in the neighbourhood (especially not the predators feeding on the pests), and one that disappears from the environment quickly, leaving no harmful by-products. This means choosing a product designed quite specifically for the particular pest you are dealing with. Avoid general products that are poisonous to a range of species.

Do not allow any pesticide, even one that is very specific and completely non-toxic to mammals, to contaminate your garden pond or a stream or river. Even the 'safe' organic sprays are often very poisonous to fish.

Scarers Birds can be pests, but you cannot deliberately poison them or, in a private garden, shoot them. You can scare or deter them, and this is fairly effective. You can use black thread stretched over the ground to form a mesh. This technique can be improved by washing metal milk-bottle tops until they are clean and shining, then threading them on lengths of black button-thread or twine. The thread must be difficult to see, but strong enough not to break in the wind. Using sticks as supports, arrange the lines of milk-bottle tops to form a grid about 30 centimetres above the tops of the emerging crops. The tops will wave in the wind, make a slight noise, and present birds with what looks to them like an impossibly difficult landing field.

You can go further and use empty plastic soft drinks bottles. Paint them so they cease to be transparent, then make cuts in the sides to you can fold out about four flaps around each bottle. Mount the bottles upside down on canes at strategic intervals among your crop so they stand about 30 centimetres above the plants. The flaps will make them spin in the wind, like toy wind-mills but much cheaper.

The next step after that is a full-

blown scarecrow, and if that does not work you will have to employ someone to run up and down the rows all day beating a drum (believe it or not, in some parts of the world children are employed to do just this).

Soap and sand The old-fashioned way to deal with aphids was to spray the affected plant with soapy water. The technique went out of fashion when the chemical companies convinced everyone that they were better off spending their hard-earned cash on products with impressive names that were much more poisonous — there were no marks for insecticides called 'suds'.

Soap does have genuine disadvantages. Hard soaps (the kind sold as bars or tablets), as well as soap flakes and powders, are made using sodium compounds which make them poisonous to plants. Liquid and jelly soaps, on the other hand, contain potassium compounds, and they actually nourish plants. Canadian scientists have now revived the old idea by developing a potassium-based soap insecticide. You dilute it before spraying and it sticks to plants, so insects that were missed by the spray tend to be coated when they walk across it. It penetrates their bodies, causing dehydration, and it contains fats and oils that disrupt the nervous system. The insecticide is harmful only to insects and breaks down within a day, leaving no poisonous traces, but it can injure some sensitive plants.

The other way to damage tough insects such as beetles involves breaking through their outer skin — the exoskeleton — so they lose body fluids. You can do this with 'diatomaceous earth', a kind of sand made by grinding to a fine powder the remains of diatoms, single-celled aquatic plants that form protective shells made from silica. The powder is so harmless to vertebrates that you could eat buckets of it without coming to harm, but while it remains dry it has an abrasive effect on passing insects, wearing away their skeletons like sandpaper. It will not deteriorate providing you keep it dry, and it will go on working for as long as it remains where the insects have to walk through it.

Biological pest control Herbivorous animals are preyed upon by carnivores. This rule applies among small invertebrates just as much as it does among large herds of game and packs of wolves. In theory, therefore, it should be possible for populations of pests to be kept in check by their natural predators. This is one approach to biological pest control. It works in some situations, but it calls for an intimate knowledge of the way of life of all the animals concerned. If the predator eats too many of the prey, for example, it may then starve, allowing the pest to return, or turning to other sources of food and becoming a pest in its own right. The predator may eat beneficial species as well as, or even in preference to, the pest. Thus the pest species should never be destroyed altogether, and food must be available for it so that

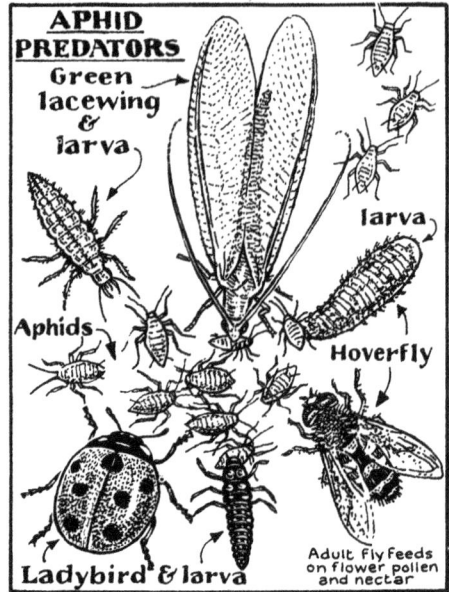

APHID PREDATORS Green lacewing & larva. Larva. Aphids. Hoverfly. Ladybird & larva. Adult Fly feeds on flower pollen and nectar

a balanced population of pests and predators is established.

An alternative approach is to infect pest populations with diseases to which only they are susceptible. Even parasites suffer from parasites (a parasite whose host is another parasite is called a hyperparasite). This too requires detailed knowledge of the species involved, and can be very effective.

If you decide to try a biological approach to pest control you will have to learn as much as you can about the pest species and the entire community of species to which it belongs. If this seems too difficult, remember that all animals can be sick and they all have enemies. Make sure that whatever controls you use do not injure the species that are trying to help you.

Sterile-male technique Some pests are present all the time, and their numbers change little. Others, however, are opportunists, and it is these that cause major infestations. When food is plentiful they multiply rapidly to take advantage of it. This 'population explosion' can be avoided if the pest can be prevented from reproducing. Provided the pests mate to reproduce (and not all do), provided the female mates only once, and provided the females do not migrate a long distance prior to mating, it is possible to disrupt the breeding pattern using a highly technical method. Insects are collected, bred in captivity, and the males collected; it is necessary to accumulate a very large number of males. They are then sterilized using chemicals or, more usually, by irradiating them. Then, just when the females in the wild are ready to mate, the males are released. Females mating with sterile males will lay unfertilised eggs that fail to develop. This technique has been used successfully with a range of insect pests.

Insecticides

The early 'first generation' insecticides were simple poisons, many based on arsenic, which were very poisonous to mammals including humans, as well as to insects. These were replaced by organic compounds, the 'second generation' of which were mainly long-lasting organochlorines, hydrocarbons (compounds of hydrogen and carbon) with chlorine added. These were effective, but insects acquired resistance to them, and although many of them were of very low toxicity to mammals, they accumulated along food chains and caused damage to other 'non-target' species, especially birds of prey. They were replaced by 'third generation' compounds, which break down rapidly in the soil and so have fewer environmental effects. Some of them, however, are extremely poisonous to mammals. 'Anticholinesterase' compounds are similar in their action to nerve gases. They inhibit the action of cholinesterase, the enzyme that

breaks down acetylcholine and so cancels chemical messages to muscles, causing convulsions, paralysis and death. Insecticides may be 'non-systemic' or 'systemic'. Non-systemic compounds act directly on the pest. Systemic compounds are taken up by plants, spread throughout the plant tissues, and without affecting the plant render it poisonous to insects.

Botanical Groups of compounds extracted from plants, or synthesized to mimic plant products, including derris (or rotenone), made from various leguminous plants and used against aphids, thrips, lice and mites; pyrethrins, made from a tropical species of *Chrysanthemum* and used against a wide range of insects; and nicotine, made from tobacco plants and used against aphids and other insects. Derris is harmful to fish. Pyrethrins do little harm to mammals or other wildlife. Nicotine is poisonous to mammals, including humans, but loses its toxicity rapidly after application. Synthetic pyrethroids, such as permethrin, are not poisonous to mammals, but can harm other insects including bees and may be extremely dangerous to fish.

Carbamates Group of anticholinesterase compounds including carbaryl (used to kill earthworms in turf as well as insects) and methomyl (used mainly against aphids), both of which are non-systemic. Systemic carbamates include aldicarb, which kills nematodes as well as insects, and carbofuran. Despite being anticholinesterase compounds, carbamates are not especially poisonous to vertebrates, and are short-lived.

Organochlorides Group that includes DDT and lindane (HCH). They attack the nervous system of insects, and in very large doses some of them affect the nervous system of mammals. They are very persistent, and use of them is being reduced. Those that are dangerous to mammals have either been withdrawn completely or are restricted in use. DDT is no longer used.

Organophosphates Group of anticholinesterase compounds, varying widely in their toxicity to mammals, including azinphos, dichlorvos, malathion and fenitrothion, all of which are non-systemic. They are short-lived, but while they last are harmful to mammals. The systemic group includes demeton-S-methyl, dimethoate and phosphamidon, used mainly to control aphids and mites, and poisonous to vertebrates.

Herbicides

Several groups of chemical compounds are used to kill weeds. Each group has its own characteristic method of operation. The list below describes the main groups. Some act on contact, others are 'translocated' (absorbed through the roots and spread throughout the plant).

Aliphatics Group including TCA and dalapon, used to kill grasses, including couch, either by damaging the roots or by translocation.

Ammonium compounds Group that includes difenzoquat, diquat and paraquat, of contact or translocated compounds used to kill a wide range of mainly broadleaved weeds. They can also be used on

crops themselves to remove leaves or kill the haulms.

Benzonitriles Group including bromoxynil, chlorthiamid and ioxynil, used mainly to control broadleaved weeds among cereals. They kill on contact, and are applied either to the foliage or the soil, where they destroy roots.

Carbamates Large group of short-lasting chemicals including asulam, barban, propham and triallate, which either kill germinating weed seeds in the soil, or are translocated. They are used for general weed control.

Diazines Group including bromacil and lenacil, of chemicals applied to the soil to kill germinating seeds and roots of a range of weeds.

Hormone weedkillers Group including MCPA, 2,4-D and 2,4,5-T, which mimic plant hormones and over-stimulate growth, used mainly as defoliants.

Inorganics Simple compounds such as sodium chlorate, used for total weed control (for example on paths), or on crops to kill the haulms. Sodium chlorate is highly inflammable and is usually mixed with an additive to reduce the risk of fire.

Phenolics Group including DNOC (dinitro-ortho-cresol), DNBP (dinoseb acetate), bromofenoxin and nitrofen, used mainly to kill weeds among peas, beans and other leguminous crops. Some, such as dinoseb, are also used on crops themselves to kill the haulms.

Triazines Group including atrazine and simazine, of chemicals that destroy the germinating seeds and roots of grasses.

Ureas Group with names ending in 'uron', including diuron, linuron and monolinuron, used to control all weeds on land not intended for cropping (such as paths), or to kill germinating seeds in the soil.

Fungicides

In the 1840s, when the fungus *Phytophthora infestans* caused late blight of potatoes throughout Britain, potatoes growing downwind of a copper refinery in Wales were not affected, and people realized that copper is poisonous to most fungi. It is used to this day. So are compounds of tin and mercury, and some of them are extremely poisonous to vertebrates. Seed dressed with mercury-based fungicide has killed people who ate it by mistake. The use of mercury fungicides is now very restricted indeed, and you will not use them in the garden.

Dithiocarbamates Group including cufraneb, mancozeb (which also contains zinc), nabam, propineb, thiram and zineb, many of them with names ending in 'neb'. They are of very low toxicity to mammals.

Inorganic Group of simple compounds including Bordeaux mixture (copper sulphate, calcium oxide and water) and other copper compounds, most of which have 'cop' or 'cup' in their names; and sulphur compounds. Copper compounds are poisonous to vertebrates, and sulphur compounds may taint produce.

Organo-tin Group including fentin (hydroxide or acetate), which are poisonous to vertebrates and very poisonous to humans (with no known antidote) if they enter the bloodstream (but not if swallowed or placed on unbroken skin).

Phthalimides Group including captan, which is harmful to fish but of low toxicity to mammals.

Pesticide poisoning symptoms

A victim of poisoning may suffer from any of these symptoms:
- Anxiety (feeling of apprehension)
- Burn marks or blisters on the skin
- Confusion
- Convulsions
- Cramp
- Diarrhoea
- Dizziness
- Excitement
- Excessive salivation and metallic taste in the mouth
- Face very pale or bluish in colour
- Fatigue
- Headache
- Inflammation around the eyes
- Mouth and throat sore, difficulty in swallowing
- Nausea
- Staining of the skin
- Sweating
- Thirst
- Trembling or twitching of muscles
- Vomiting
- Watering of the eyes

Concentration through food chains
Organochlorine compounds are chemically stable. When they were introduced their stability was one of their principal advantages. It meant they did not break down rapidly, and once applied they went on killing pests for a long time, reducing the need for further applications.

But the stability that seemed an advantage turned out to be double-edged. They were not highly specific, and harmed many species other than those against which they were directed. Their stability meant that these adverse effects were long-term, and in particular they could be concentrated along 'food chains'.

Organochlorine compounds are insoluble in water, but dissolve in fats and oils. When an animal eats something coated with an organochlorine, the compound may find its way into body fats and remain there. If a predator eats a number of prey animals, each of which has organochlorines stored in its body fat, the organochlorines are collected; if contaminated predators are then eaten by carnivores further along the food chain, organochlorines from each of them are stored. Eventually, concentrations may reach levels at which they cause harm, especially in birds, where they can for example interfere with the metabolism of calcium, causing the females to lay eggs with shells so thin that they crack. It was this effect, mainly on wild birds of prey, that led to severe restrictions on the use of organchlorines and the banning of many of them.

A few organochlorines are poisonous to humans, but most of them — including DDT — are harmless. Although most human fat (and milk) contains

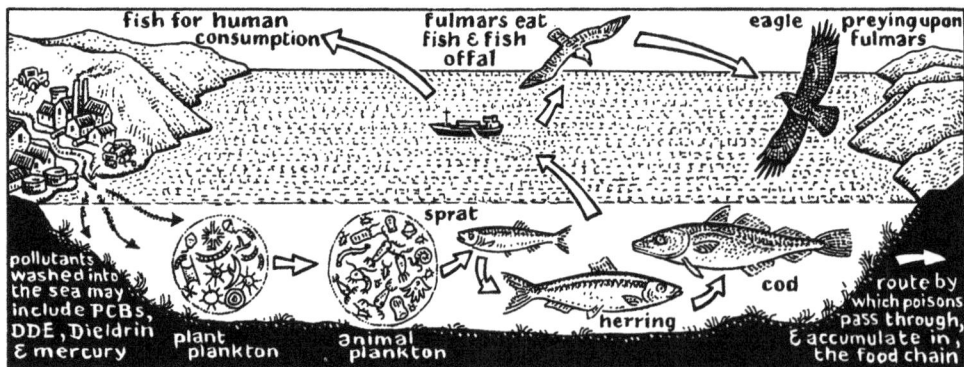

fish for human consumption; fulmars eat fish & fish offal; eagle preying upon fulmars; sprat; cod; herring; pollutants washed into the sea may include PCBs, DDE, Dieldrin & mercury; plant plankton; animal plankton; route by which poisons pass through, & accumulate in, the food chain

some organochlorine residues, surpluses are released and excreted before concentrations reach harmful levels.

Livestock

Pets that live outdoors are called livestock, but if you keep only a few of them they will be pets all the same. You can keep some even in a small garden. Chickens are easy, and so are bees if you take a few precautions, but they are no more than a beginning. Perhaps you could keep a few pigeons or ducks or rabbits. You might keep a pig or a goat, or raise a few turkeys for Christmas.

Why should you? Environmentally keeping a few animals is quite sound. Chickens and pigs, for example, recycle food scraps and spoiled food, converting it partly into food for you but also into rich manure for the garden. They cause no pollution while they process it and save you the trouble of composting wastes. If you have children, and especially if you live in a town with only limited access to the farmed countryside, the educational benefit to them is great. Learning to care for animals a little different from household pets will increase their self-confidence and their awareness of the needs of non-humans. That will make them more appreciative of wildlife and its needs. By watching the animals they will learn much about animal behaviour, and in particular about the social organization of other species. This may help them to recognize that cooperation and altruism are far more important in nature than competition and aggression.

You can always keep 'farmyard' animals as well as pets, of course, though you may need to protect some of your stock from the unwelcome attentions of predatory cats and dogs.

In recent years some people have come to frown upon the keeping of pets. They maintain that pets eat food that might otherwise feed humans. This is untrue: petfoods are made from food wastes or food that has been classified as unfit for human consumption. It is also argued that pets represent a health hazard for humans, that they harbour transmissible diseases. This is true, but largely irrelevant. Some people are allergic to fur, feathers, or substances in the saliva of cats or dogs, and they cannot keep animals, but in practice the transmissible diseases are either fairly mild in humans, or transmitted so rarely as to be of little importance. In any case, the diseases themselves are caused by parasites that are quite easy to control.

Diseases transferred from pets to humans The risk of infecting humans with dire diseases is often exaggerated. In the tropics there are real risks, but in Britain they are small.

Roundworms excreted by puppies can infest humans (*see* page 96), but the dangers are easily avoided. One species of tapeworm can also be passed from dogs to humans, but only from dogs that acquired it in the first place from infested sheep. Children are the most likely victims of both worms, because they play on the ground and may come into direct contact with faeces.

Psittacosis is a disease affecting the lungs that can be transmitted by birds to many mammals, including humans.

Apart from rabies, which you will not catch in Britain, there are no other serious illnesses you can catch from your pets, though intestinal bacteria can cause illness.

If your pets are cared for properly they will not suffer pest infestations. Make sure they receive regular veterinary checks and immunization. Discourage dogs from licking human skin, and children from encouraging them to do so. Do not allow pets to touch food you intend to eat, observe ordinary rules of hygiene, and you and your pets will remain healthy.

Roundworms Do make sure that your pet cats and dogs are treated regularly for intestinal worms. Your veterinary surgeon will advise you, and provide a record card showing the date and type of treatment the animal has received, and reminding you of the date when the next treatment is due.

Roundworms are rather different. The best known is called *Toxocara canis*, though there is also a feline species. For most of the time they remain dormant, but when a female becomes pregnant hormonal changes in her body activate the worms, which can be transferred to the foetus so the young are born infested. For a short time while they are very young, kittens and puppies will then pass the worms in their stools. If the stools fall in a place where conditions are favourable to the parasites, the eggs may germinate after a few weeks, and if swallowed can infect new hosts.

In most cases, once the parasites are inside the human body they are destroyed by the body's immune system, but if that line of defence fails the worms may migrate, occasionally reaching the eye. Wherever they lodge they will form temporary cysts; children have been blinded, but in most cases because of the surgical intervention that was unnecessary — left to themselves the cysts would have disappeared. Such infection is extremely rare, and the danger from *Toxocara*

canis much exaggerated.

The cycle can be broken very easily if you treat the female to kill the worms early in the pregnancy, and again shortly after she gives birth.

Keeping 'farmyard' birds

Chickens Provided there is no local regulation or byelaw to prevent it, it is easy to keep chickens in a small space. Fence an area in which they can scratch, take dust-baths and shelter in hot weather; provide a hen-house where they sleep with nesting boxes from which you will collect eggs, and apart from food and water they need nothing more.

Feed them on bought corn augmented with calcium (their own eggshells baked in the oven and crushed into their food), scraps of meat or cheese from the kitchen, and cabbage or other brassica leaves; or boil up all the kitchen scraps to make a mash. You can use artificial lighting to cheat them into laying eggs throughout the winter, otherwise they will cease laying when the days grow shorter than a critical length, to start again as the days lengthen past that critical limit.

You will have to arrange for someone else to feed them while you are away from home, but otherwise they require very little attention.

Ducks It is possible to keep a few ducks in the garden if you have no pond, but they love water and it seems cruel to deny it to them. They require less elaborate accommodation than chickens because they do not perch, but they do fly, so you need to clip their wings if you are keeping them in a small space. Discourage predators by clearing away most of the low vegetation, especially near water.

They will find a good deal of their own food, consisting of grass, slugs, snails and insects, but you will have to give them some extra corn and cooked offal — not too much or they will lay down fat. They are poor mothers, so you will need an incubator or broody hen to hatch the eggs, and will have to keep the ducklings away from water until they are about a month old.

Geese Geese are committed vegetarians, and live happily on nothing but grass, but it must be kept short. If you want them to grow fast, though, you will have to give them extra corn. Traditionally they are kept in orchards, and need plenty of space to exercise since they are big birds. You have to look after them carefully while they are very young, but they are hardier than ducklings and by the time they are a month old they have passed the danger period. They can live for nearly twenty years.

Geese are excellent guardians and will drive away a solitary dog (though not a pack) and most humans.

Guineafowl Originally from Africa, guineafowl have lived in Britain for a long time as more or less domesticated gamebirds, kept for their meat. They fly well enough to roost in trees, but otherwise leave the ground reluctantly. When danger threatens, they run. They are gregarious and extremely noisy, so although they are not aggressive they give ample warning of intruders. They

need space in which to wander, but otherwise largely look after themselves.

Guineafowl feed on insects and other invertebrates, seeds, leaves and bulbs.

Pheasants and peafowl Peafowl (the peacock is the male) are one species of pheasant, and like all pheasants were raised originally for their meat, though to us their main function is decorative. They are difficult to keep in a confined space and are very combative, but if you have a small patch of scrub or woodland with trees for them to roost in, they will thrive.

Provide them with a house with its back to the prevailing wind, where they can perch on really cold and wet nights. They eat grain and grass and will find most of their own food, but a little extra corn will be much appreciated.

Pigeons Many people keep racing pigeons, but at one time pigeons were kept in dovecots mainly for food. A dovecot is a kind of tower with nesting boxes covering the inside wall and openings through which the birds can leave and return at will. Feed them grain which will make sure they return home after their flights, allow them to mate, and take the young birds ('squabs') for the table.

Turkeys You can raise your own Christmas turkeys in a very small shed or lean-to barn, allowing about five square metres for each bird. They are difficult to breed, but you can buy the stock as poults in mid-summer. After that you must feed them a proprietory feed as they are unlikely to accept anything else. They are delicate when young, but once they have their red 'caruncles' around their throats they are past the danger period and are henceforth hardy.

'Farmyard' mammals

Goats Unless you have plenty of time to spend collecting its food, a goat needs about one-fifth of a hectare of land on which to forage. Keep it tethered, though, or you may never see it again — goats can jump or climb almost anything. They will climb trees if there are branches low enough to get them started.

A goat also needs housing at night and for most of the winter, with a rack to hold its food. Some goat breeds will eat grass, but most are browsers, preferring leaves. They will eat gorse and heather, and if there is a food processing factory nearby the goat will appreciate any wastes they can offer. If you want the goat to produce milk you will have to supplement the food it finds for itself. Remember, though, that although a goat will eat almost anything, it will not yield much milk unless it is well-nourished; if you feed it strong-tasting or smelly garbage the milk will be tainted. The goat is really a smallholder's animal, but there may be a patch of waste ground near you that would suit a goat well. There are many books on goat-keeping.

Pigs Not so many years ago every rural family had its pig. It would be bought as a weaner, housed in a sty with a bit of yard in front (pigs like being in the open air), and fed scraps

collar loin corner gammon
thick back prime back
flank
thin streaky
forehock thick streaky gammon hock

from the kitchen, plate scrapings, weeds from the garden, windfall or rotten apples, and anything else that came to hand. Its manure would be used as fertilizer, and sometimes the pig might be allowed its freedom to dig the garden, which it did very thoroughly in its search for edible roots. When it was big enough it was killed and the family had pork, bacon and ham to keep them going for months. There are many books on this kind of backyard pig-keeping and it is not difficult, provided you have a backyard and the local authority and neighbours do not object.

Rabbits The traditional source of meat for the rural poor, rabbits are very easy to keep in hutches, and you can stack the hutches like tenement blocks. You will have to decide which breeds to keep, but after that it is simple. Workers used to take a bag to work with them and fill it up on the way home with vegetation from the roadside to provide supper for their rabbits. Use straw for their bedding and compost it when the rabbits have finished with it.

Insects

Bees Before you decide to keep bees, you must make sure that neither you nor any member of the family or close neighbours are allergic to bee venom. A bee sting is no more than mildly irritating to most people, but anyone who reacts badly to it can become seriously ill or even die.

Then you need to contact your local beekeepers' society or club, where you will find people eager to explain the intricacies of their craft and to help and supervise you. They will also guide you in the direction of the equipment and stock you will need, and warn you if your area is already fully populated and has no more room for bees.

Once your hive is installed and stocked, however, beekeeping is not difficult and takes up very little time.

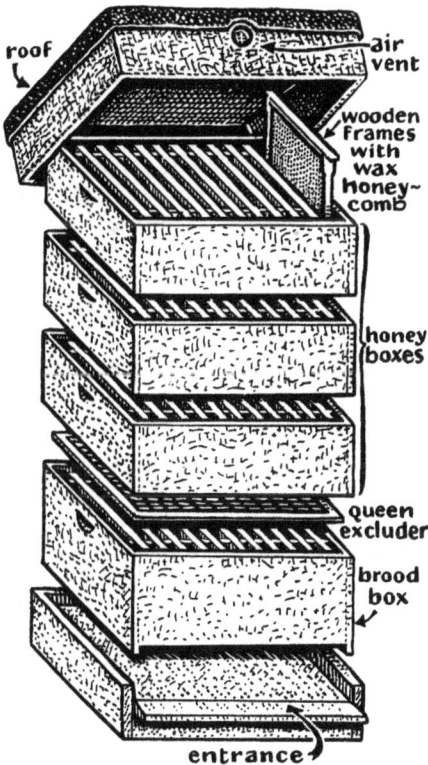

Non-edible plants

No matter how keen you may be on the fruit and veg, most of the garden will be given over to decorative plants you cannot eat, but just because you cannot eat them it does not follow that nothing else can. So how can you make the parts of the garden that you find inedible into an attractive place for the non-human residents and visitors?

Hedges Britain was once forested. Most of our native plants and animals are woodland species, especially species of the woodland edge, where sunshine falls directly on to the ground and allows the herbs we know as spring flowers to flourish. A hedge can mimic the edge of woodland, and for this reason hedges are of major importance to our wildlife. There is hardly a native plant or animal that cannot be found in some hedgerow or other. Hedgerows also form pathways, routes by which animals can move around the countryside without departing from their source of food, shelter and security; but they can only be pathways if they are linked one to another.

If you look at the layout of the gardens in your immediate neighbourhood you should be able to see how their boundaries form, or could form with a bit more planting, miles of continuous hedgerow.

Any hedge made from plants will help wildlife, but the best is one made from hawthorn interspersed with native broadleaved trees that are allowed to grow to their full size.

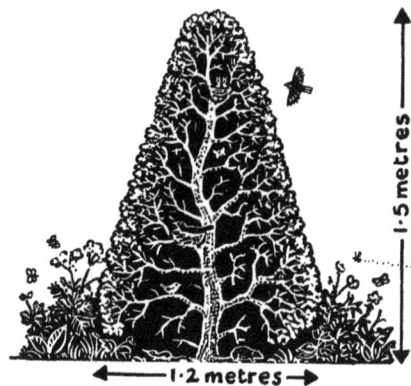

Design the hedge so that it is the shape of a capital letter A in cross-section. Pigeons and some other birds will not nest below 1.5 metres, so if you want to attract them some parts of the hedge must be tall (an A-shaped hedge 1.5 metres tall will be about 1.2 metres thick at the base). Allow other plants to fill in the base; they will provide dense cover and a fairly constant climate, sheltered from wind and rain, for anything living inside the hedge.

Ivy A native British plant that provides food and shelter to many small animals, ivy is a climber but not a parasite. It obtains all its nutrients from the ground in which it is rooted, and uses walls and trees only for support. It anchors itself to them, but does not interfere with them in any other way. It can injure trees, but only if it grows so prolifically that its weight breaks branches. If a living tree is festooned really thickly with ivy, remove most of it.

Ivy will not damage garden walls, but unfortunately it will damage house walls because it will trap moisture and allow it to penetrate. Allow the ivy to grow on garden walls, then, but remove it from the house or you could end up with a large repair bill.

Lawns Most gardens have a lawn, and if you have young children you will need somewhere for them to play, but the fact remains that a perfectly manicured lawn takes a great deal of management. It is also very boring, and with less management it can become an interesting place. Mow the grass when necessary, but not too short; throw out stale bread on to it when the ground is hard, and spend the rest of your time watching it.

Notice how quickly dead leaves and old grass cuttings disappear from the surface. They are eaten or dragged below the surface by small animals, mainly earthworms.

Because it is open, birds are able to feed on it safely. They can see predators while they are still far enough away to make escape easy. The birds feed on the many small animals they find, and if you look hard you will find some of them too. With a little less effort on your part, the lawn might become a haven for wildlife.

Lawn mowing If you never mowed the lawn, in time it would cease to be a lawn at all. The grass would grow tall, and then shrubs and eventually trees would appear. As they grew they would shade out the grass — most of Britain would revert to woodland if we gave it a chance. So you have to mow the lawn, but grass is unique among plants in that cutting it actually encourages it to grow — it does it good. Try to mow it frequently so there are never many grass cuttings at a mowing, and leave the cut grass where it falls to be recycled by the soil animals and bacteria. Do not bury the lawn deep in cuttings, because that will injure the grass.

Plants to encourage wildlife Apart from the lawn, which is inevitably an open area, and the shaded ground beneath trees, maintain a fairly dense cover of plants throughout the year. Do not be over-enthusiastic with your weeding. Leave a few brambles, thistles and other uninvited plants, and even a little bindweed.

Plants with bright flowers attract insects, but encourage as many different colours and flower shapes as you can to cater for nectar-feeders, some of which visit only certain plants. Sage and foxgloves, for instance, are bumble-bee plants. If you shine ultra-violet light on a foxglove flower you will see it the way a bee sees it, with a guide path marked out clearly on its petals. Buttercups, on the other hand, attract many species of beetles, moths, butterflies and other insects, some

Pollen and nectar advertising

foxglove

bees land here

pansy

lines point to pollen source

evening primrose

As we see it, with plain yellow petals..

..and as a bee would see it, with an ultra-violet target.

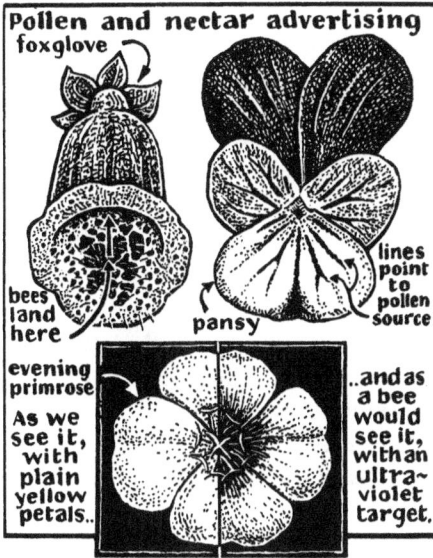

calling for nectar, others for pollen. If you grow leaf vegetables, leave a few of them to flower.

Think of the different kinds of habitat that exist in nature, and try to reproduce them on a small scale. Allow some areas to be exposed to direct sunshine, with others in shade. Let your rockery provide exposed stones on which, if you are lucky, 'cold-blooded' vertebrates may bask in summer, and crevices beneath the stones where they can shelter during the midday heat. If you provide small areas of many different habitats, and plants to provide shelter and food for herbivores, your garden will attract many species of animals.

A pond will attract frogs, toads, or even newts — it need not be elaborate, and should not be too deep. It can be lined with heavy duty polythene weighted with rocks round the edges.

Trees All trees are helpful to wildlife, but the most helpful are those that have grown here for thousands of years. Communities of animals and plants develop together, and while there are exceptions, the richest relationships are generally those that have been estab-

lished longest.

Fruit trees are the exception. They will supply the household, and damaged fruit will feed countless invertebrates and birds.

Do not plant trees closer to a building than about nine metres, because they cast too much shade and their roots can undermine foundations. Every garden, however, should have a few trees.

Wildlife

Conservation consists mainly in limiting the damage humans cause to wild plants and animals, and there are two aspects to it. The first is to modify our own activities so that we cause less interference — we might call this 'passive conservation'. The second is 'active conservation', and it involves improving natural habitats and creating habitats where before there were none. Your garden is — or could be — an important habitat area. It is under your own control, and so is an obvious place for your own active conservation effort to begin.

Not many years ago naturalists were contemptuous of gardens. They were full of exotic plants introduced from goodness knows where and then hybridized until they were as artificial as a high street hamburger. Gardens were multicoloured deserts, manicured and managed by gardeners who were little more than exterior decorators.

Then two things were noticed. The first was that despite their artificiality gardens seemed to attract birds, butterflies, bees and many other small animals — and occasionally larger ones. The second was that when you add together the total area of gardens in the average suburban neighbourhood, the result is an impressively large area.

Today gardens are seen more accurately as refuges for wildlife. What animals can you find in your garden, and how can you attract more?

Invertebrates

Butterflies and moths Britain is close to the northern geographical limit for butterflies and moths, which is why their numbers fluctuate between 'good' and 'bad' years. The careless use of insecticides has some effect, but the weather is a much more important influence.

You can buy eggs and raise your own insects, in which case the supplier should tell you the food plants they require, but it is much better to provide suitable food and conditions and attract native species in that way. Grow trees, and plants with bright scented flowers, and provide a mixture of shade and places exposed to bright sunlight. These will attract adults to feed and rest. If you also provide food plants for caterpillars, the adults will breed in your garden. Attract adults and allow them to breed, and within a year or two your garden will be a haven for them and a delight for you.

Butterflies fly by day, and prefer bright sunshine and little wind. Some moths fly by day, but most are nocturnal. If you want to watch them you will have to wait for a warm dark summer night with little wind.

Food plants of
common butterflies and moths

beech: pale tussock moth (*Dasychira pudibunda*), tau emperor moth (*Aglia tau*), barred sallow moth (*Xanthia aurago*)

bindweed: convolvulus hawk-moth (*Herse convolvuli*)

birch: (also alder) alder kitten moth (*Harpyia bicuspis*), Kentish glory moth (*Endromis versicolora*), pebble hook-tip moth (*Drepana falcataris*), scalloped hook-tip moth (*Drepana lacertina*), orange underwing moth (*Archiearis parthenias*), large red-belted clearwing moth (*Aegeria culiciformis*); (also sweet gale and bog whortleberry) large argent and sable moth (*Rheumaptera hastata*)

blackcurrant: spinach moth (*Eulithis mellinata*)

bramble and raspberry: peach-blossom moth (*Thyatira batis*)

buckthorn: brimstone butterfly (*Gonepteryx rhamni*)

burdock: orange ear moth (*Gortyna flavago*)

buttercup: flame brocade moth (*Trigonophora flammea*)

campion and ragged robin: campion coronet moth (*Hadena rivularis*)

chervil: chimney sweeper moth (*Odexia atrata*)

crucifers: green-veined white butterfly (*Pieris napi*), orange tip butterfly (*Anthocharis cardamines*)

dock, sorrel: small copper butterfly (*Lycaena phleas*)

elder, honeysuckle, viburnum: swallow-tailed elder moth (*Ourapteryx sambucaria*)

gorse, ling, broom: green hairstreak butterfly (*Callophrys rubi*)

grasses: meadow brown butterfly (*Maniola jurtina*), ringlet butterfly (*Aphantopus hyperantus*), small heath butterfly (*Coenonympha pamphilus*), gatekeeper butterfly (*Pyronia tithonus*), (especially couch grass) speckled wood butterfly (*Pararge aegeria*), wall brown butterfly (*Lasiommata megera*), large skipper butterfly (*Ochlodes venatus*), reed tussock moth (*Laelia coenosa*), drinker moth (*Philudoria potatoria*), dark arches moth (*Apamea monoglypha*), Mother Shipton moth (*Callistege mi*), straw dot moth (*Rivula sericealis*)

hawthorn: short cloaked moth (*Nola cucullatella*), sulphur thorn moth (*Opisthograptis luteolata*)

heather: emperor moth (*Saturnia pavonia*), common heath moth (*Ematurga atomaria*)

hemp nettle: small rivulet moth (*Perizoma alchemillata*)

holly, ivy, buckthorn: holly blue butterfly (*Celastrina argiolus*)

honeysuckle, lilac, privet: lilac thorn moth (*Apeira syringaria*)

legumes, especially vetches (S. England only): clouded yellow butterfly (*Colias crocea*), common blue butterfly (*Polyommatus icarus*)

lichens: four-spotted footman moth (*Cybosia mesomella*), large footman moth (*Lithosia quadra*), rosy footman moth (*Miltochrista miniata*), scarce footman moth (*Eilema complana*), red-necked footman moth (*Atolmis rubricollis*)

lime and elm trees: lime hawk-moth (*Mimas tiliae*)

nettles: red admiral butterfly (*Vanessa atalanta*), peacock butterfly (*Inachis io*), small tortoiseshell butterfly (*Aglais urticae*), burnished brass moth (*Diachrysia chrysitis*), snout moth (*Hypena proboscidalis*); (also raspberry) scarlet tiger moth (*Callimorpha dominula*); (also thistles) painted lady butterfly (*Vanessa cardui*); (also willow and other trees) comma butterfly (*Polygonia c-album*)

oak: purple hairstreak butterfly (*Quercusia quercus*), merveille du jour moth (*Dichonia aprilina*), scarce silverline moth (*Pseudoips bicolorana*); (also birch and alder) autumn green carpet moth (*Chloroclysta miata*)

plantain, scabious: marsh fritillary butterfly (*Euphydryas aurinia*)

pine needles: pine hawk-moth (*Hyloicus pinastri*), pine processionary moth (*Thaumatopoea pinivora*), pine lappet moth (*Dendrolimus pini*) *poplar:* poplar lutestring moth (*Tethea or*), poplar hornet clearwing moth (*Sesia apiformis*)

privet: privet hawk-moth (*Sphinx ligustri*)

ragwort: cinnabar moth (*Tyria jacobaeae*)

sorrel: common forester moth (*Procris statices*)

sowthistle: shark moth (*Cucullia umbratica*)

spruce: black arches moth (*Lymantria monacha*)

violets: dark green fritillary butterfly (*Mesoacidalia aglaja*), high brown fritillary butterfly (*Fabriciana adippe*), small pearl-bordered fritillary butterfly (*Clossiana selene*), pearl-bordered fritillary butterfly (*Clossiana euphrosyne*)

willow: lappet moth (*Gastropacha quercifolia*); (also apple) eyed hawk-moth (*Smerinthus ocellata*); (also birch and poplar) pale prominent moth (*Pterostoma palpina*); (also poplar) poplar hawk-moth (*Laothoe populi*), great prominent moth (*Peridea anceps*), swallow prominent moth (*Pheosia tremula*), pebble prominent moth (*Eligmodonta zicsac*), white satin moth (*Leucoma salicis*), herald moth (*Scoliopteryx libatrix*), blue underwing moth (*Catocala fraxini*)

willowherb: large elephant hawk-moth (*Deilephila elpenor*)

Position light near a hedge that contains a wide variety of native species like hawthorn & blackthorn

powered by a small generator, but any bright light will do. Use old egg-cartons to make up 'buildings' on the sheet a little way from the lamp, so the insects can find shade and dark surfaces — then wait. If the moths are flying it should not be long before they begin to visit your sheet.

Slugs and snails You may think of slugs and snails as pests and find them revolting. The truth is that very few of them are pests, and all of them are fascinating. There are slugs and snails that hunt other slugs and snails, for example, and brave slugs that will challenge you if they feel threatened (though the challenge is of necessity a slow one, and if you call their bluff they will collapse into a self-protective dome).

Their lives are dominated by the need to avoid dessication. They are mainly nocturnal, and prefer damp places. Snails can avoid drying out by shutting themselves inside their shells; slugs lack shells, but can burrow into the ground.

Most species feed on rotting plant material, lichens, algae and fungi, and some slugs will climb high into trees to graze the lichens. Some slugs (but rarely snails) will eat cultivated vegetables because they usually have soft leaves and resemble rotting plants. You

Moth-watching Most moths fly at night, so that is when you must observe them. Choose a fine night with little wind, and wait until it is quite dark. Spread an old white sheet on the open ground, and place the brightest light you can find in the middle. You can buy lamps designed for the job and

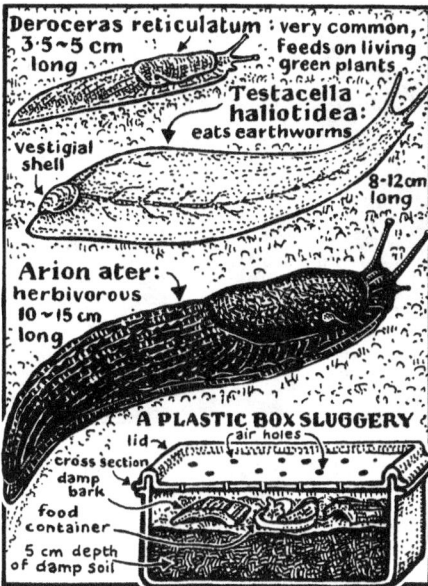

Deroceras reticulatum : very common, 3.5~5 cm long · feeds on living green plants

Testacella haliotidea : eats earthworms
vestigial shell
8-12 cm long

Arion ater: herbivorous 10~15 cm long

A PLASTIC BOX SLUGGERY
lid
air holes
cross section
damp bark
food container
5 cm depth of damp soil

can keep them in captivity where they will eat a wider variety of food, and they can digest cellulose, a convenient talent which allows them to eat paper and cardboard. They in turn are food for many birds (most notoriously the voracious song thrush), mammals (including some rodents as well as hedgehogs and shrews), certain beetles (including the larvae of the glow worm which eat nothing but snails), and the carnivorous slugs and snails.

Soil animals If you examine a spadeful of earth you should find a few earthworms. You might also see a beetle or two, or a millipede, and you might then assume that you had seen all the life the soil could offer. You could not be more mistaken — the soil teems with life. If you doubt it, look at the information in the box.

Many of the soil animals are tiny, and if you want to examine them you will need a good hand lens or a low-magnification microscope, and a device for collecting the organisms.

Use a trowel to dig out a sample of soil, remove large animals such as

earthworms, beetles and woodlice, and place the soil in a Tullgren funnel. You can improvize one of these very easily.

Use smooth clean paper to make a long steep-sided cone about ten centimetres wide at the top. Place the small end of the cone into a jar containing water or a mixture of seven parts of alcohol (surgical spirit) to three parts of water; do not wet the paper funnel. Place a sieve such as a tea strainer near the top of the funnel and put your soil sample in the sieve. Set the whole device about 30cm below a hundred watt light bulb and leave it.

The light and warmth will make the animals move downwards through the sieve, into the funnel, and from there they will fall into the jar. After about 24 hours most of the animals will have arrived in the collecting jar, but it can take three or even five days to extract every last one. The liquid will kill the animals, so if you prefer to study them alive cover the bottom of the jar with damp blotting paper instead — in this case you must make sure there are no centipedes or other carnivores or they will devour your other specimens.

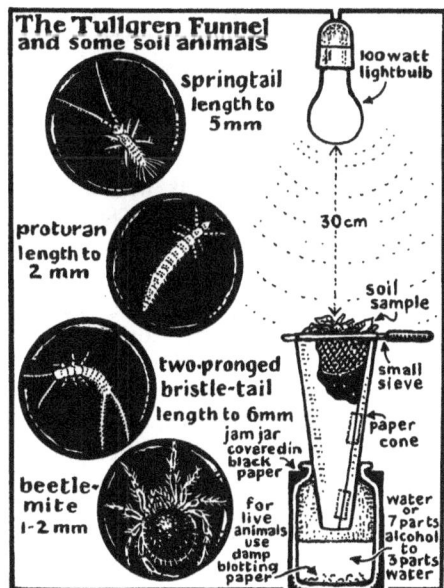

The Tullgren Funnel and some soil animals
100 watt lightbulb
springtail length to 5 mm
proturan length to 2 mm
30 cm
soil sample
two-pronged bristle-tail length to 6mm
small sieve
jam jar covered in black paper
paper cone
beetle-mite 1-2 mm
for live animals use damp blotting paper
water or 7 parts alcohol to 3 parts water

<table>
<tr><td>

Soil animals

Numbers of animals living in the top 10cm of each square metre of ordinary soil.

Numbers in thousands

Nematodes 1,800-120,000
Mites 20-120
Enchytraeid worms (potworms) 20
Springtails (*Collembola*) 10-40
Earthworms 2
Molluscs (slugs and snails) up to 8
Centipedes and millipedes 1-2
Woodlice up to 0.5
Ants up to 0.8
Fly larvae 1
Spiders up to 1
Beetles and their larvae 0.5-1

</td></tr>
</table>

Spiders Never deliberately kill a spider. For one thing it brings bad luck, for another (or perhaps the same reason phrased differently) spiders feed on the insects that might otherwise become pests. There are about 30,000 species of spiders in the world. You will not see this many in your garden, but you will see some and evidence of more, for they outnumber humans many times over. You will see beautiful orb webs, glistening in the morning dew, and sheet webs slung horizontally, their owners hidden and waiting for prey to tread or fall on to the trap. If you see a tight mass of spider silk with a denser mass inside it, it will be an egg sac, and early in the year it is likely to be filled with eggs. Many female spiders care for their young, carrying their egg sacs with them, cutting them open when it is time for the spiderlings to emerge, and then carrying their young on their backs. Wolf spiders, for example, do this: they are the small spiders you see running about on the ground, and around July or August you can see them with their offspring. Wolf spiders do not make webs, but catch their prey by running after it.

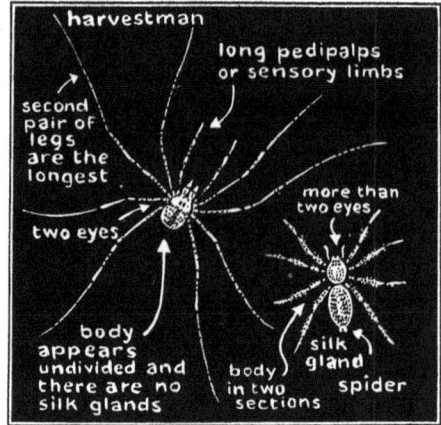

Young spiders, and sometimes older ones too, disperse from areas that are becoming overcrowded by 'ballooning'. They climb to the top of a tall plant, release a length of silk, and as it is caught by the wind so they are transported to wherever it takes them. Apart from a few common species, spiders are not easy to identify precisely, but just watching them can fill many hours with free instruction and entertainment.

Vertebrates

Fish The goldfish with which you stock your garden pond are herbivorous, and close relatives of the carp. Those that escape into rivers and lakes in southern England may survive, provided the water is not too cold, but will soon turn into silver carp-like fish. If there are ordinary carp nearby they may even breed, but if they do so all the offspring will be female. When female goldfish mate with male carp the spermatozoans are able to penetrate the outer part of the eggs sufficiently to stimulate them into dividing, but fail to reach and merge with the nucleus, so only the mother's genes are transmitted to the young.

If your garden borders a stream or river, never throw waste into it or allow garden chemicals to drain or drift into it. Fish vary in their sensitivity to

---

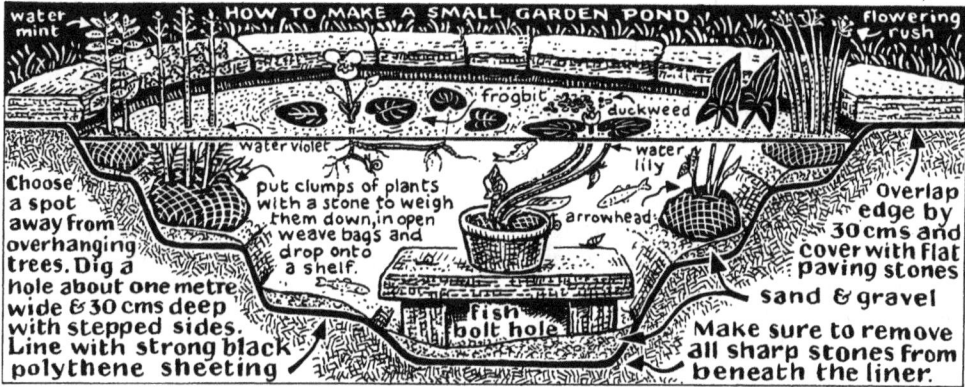

HOW TO MAKE A SMALL GARDEN POND

water mint · flowering rush · frogbit · duckweed · water violet · water lily · arrowhead

Choose a spot away from overhanging trees. Dig a hole about one metre wide & 30 cms deep with stepped sides. Line with strong black polythene sheeting

put clumps of plants with a stone to weigh them down, in open weave bags and drop onto a shelf.

fish bolt hole

Overlap edge by 30 cms and cover with flat paving stones sand & gravel **Make sure to remove all sharp stones from beneath the liner.**

pollution, but all of them are very susceptible to pesticides, and salmon and trout, the most sensitive of all, cannot tolerate anything that decreases the amount of oxygen dissolved in the water. If you see trout or salmon you can be sure the river is clean.

Eels can tolerate most things, and survive in polluted water. If you find any in your garden, do not mistake the adults for snakes or the elvers for worms. They have neither scales nor fins, are slippery and very lively, and their faces are those of fish, not reptiles. Leave them alone. They move quite long distances over land in the course of their migrations.

Amphibians

Frogs, toads and newts Frogspawn is laid in shapeless masses, toad spawn in strings. Newts lay their eggs singly, attached to stones or plants. If you find spawn or eggs in a pond or ditch, do not try taking it home to your garden pond — the animals are very unlikely to survive. If your pond suits them, sooner or later they will find their own way to it. Once they do, they will take care of themselves, and if you are lucky their numbers will increase. They tend to return to the same mating sites year after year, but some of them wander off in search of food at other times, and they spend a good deal of time settled

into the mud on the bottom of the pond.

Tadpoles are hunted by several aquatic predators. Adult toads, which move further from water than frogs, protect themselves against predatory mammals by secreting a foul-tasting slime, but it does not work with snakes, which are less fussy. Frogs too are hunted by snakes, and also by some domestic cats (and even dogs). Cats that learn to hook frogs from the pond with a paw will find the same tchnique works quite well with half-tame and unsuspecting fish. Cats and dogs rarely eat frogs, but they may kill them.

All frogs, toads and newts are carnivorous. Frogs and toads eat insects; newts vary this diet with slugs, snails, and various aquatic invertebrates.

Frogspawn · newt egg · toadspawn

Reptiles

Lizards and snakes Britain has few reptiles, but it is not impossible to meet one in the garden. The small lizard you may see basking on a stone in summer will be the common or viviparous lizard, which feeds mainly on insects. The slender slow-worm, pale brown or even pinkish in colour, is one of the many lizards that have lost their legs in the course of their evolution. It looks like a snake, but is not. It feeds on insects, slugs, snails and similar small animals, and will not harm you. If you catch it by the tail, the tail may come off in your hand as a defence mechanism — the slow-worm will grow a new tail, but seldom as neat and handsome as the one it lost.

Slow-worms

The grass snake can grow up to two metres long, though it is usually smaller. It is really an aquatic snake that spends most of its time not far from water, and it feeds on small animals. It is harmless. You can identify it by the yellowish crescent-shaped mark behind the head, and by the round pupils of its eyes.

The adder is much smaller, also feeds on small animals, has a pattern of dark blotches along its back (often in a more or less zigzag shape), its eyes have vertical pupils, and it is venomous. It will enter water, but it is not aquatic. You may find it in dry heathy areas, but it sometimes comes into nearby gardens. It does not like humans very much and tries to keep out of their way. If you should meet one, remember that it is not looking for trouble, but will defend itself if attacked — treading on it counts as an attack.

Birds

If your garden has an open exposed lawn, birds will feed happily, safely and efficiently on the worms and other invertebrates they find there. The size of the open area will determine the size of the birds that land. Large birds will not risk finding themselves grounded and trapped on a tiny lawn surrounded by high vegetation or walls.

Leave some of your garden plants to form seed heads for the benefit of seed-eating birds, and grow rowan or other trees and shrubs that produce brightly coloured berries — they too will attract birds. A bird table will attract smaller visitors, and if you hang beneath it a net bag filled with birdseed, or a head of millet, the smaller seed-eaters will have a regular supply of food.

Birds also need shelter. Encourage areas of dense vegetation and allow some of the trees and taller shrubs to grow unchecked. The vegetation will also provide nesting sites for many birds.

Swallows, house martins and swifts feed on insects caught in flight, never land on the ground, and nest in gaps in the eaves or build mud nests against fascia boards.

Birds that build nests in enclosed places with a small entrance hole can be provided with nesting boxes, but beware of overpopulating the area with bluetits and great tits. They are really woodland birds, do not travel far, and their numbers are limited naturally by the availability of nesting sites. If you provide them with more places to nest, their numbers may increase beyond the capacity of the immediate area to feed all their young, so many of the nestlings will starve.

If you find a nestling that has fallen from the nest, leave it alone. Its parents will continue to feed it and protect it as best they can, but if you move it they may not be able to find it, and if you handle it they may abandon it.

No bird enjoys being handled, so never pick a bird up unless it is absolutely necessary. If you do, wear thick gloves if it is bigger than a blackbird, because its bill can inflict a nasty wound. Do not put its head near your face because it may attack your eyes. Hold it on its back and prevent it from flapping its wings.

Mammals

Badgers If you live fairly close to an area that has been broadleaved woodland for many years, and if your garden has a large old lawn where plenty of earthworms surface at night, you may attract a foraging badger. You cannot mistake it for any other animal, especially if you get a glimpse of its striped face. If you see one, stay very still and quiet or you will frighten it. Its eyesight is poor, but its hearing and sense of smell are acute.

Badgers prefer to live in broadleaved woodland, but they patrol along regular paths in search of food, so if your garden is inside its range your badger will return to it.

They will eat most things, but are inordinately fond of earthworms. You are most likely to see a badger in the garden in autumn, for that is when they go on their longest foraging expeditions, but you could see one at any time of year. Badgers spend a lot of time

badger

sleeping in winter, but they do not hibernate and emerge from time to time to look for food.

There is nothing you can do to encourage them except leave gaps in the bottom of the hedge for them to enter. Your local trust for nature conservation will tell you whether badgers live in your neighbourhood.

Bats Bats are now the most strictly protected animals in Britain. You may not injure or disturb them in any way, and unless you have a licence or are helping a sick or injured bat that includes handling and keeping them. You may not even photograph a bat unless it is flying in the open air.

If bats roost in the roof space of your house or in any other building you own you must not block their entry to the roost. If you do the fine is £2,000 *per bat*, and a roost may accommodate more than a hundred individuals! This sounds a large number, but bats are tiny and unless you are an expert you could have difficulty finding a roost this size in your roof space — it would look like a small bundle of dry leaves hanging in a dark corner.

Obviously such strict protection may frighten some people into taking matters into their own hands and destroying bat roosts in the hope that their offence will go undetected — which it almost certainly will. It is, however, possible to have a bat roost removed legally. Contact the nearest branch of your local trust for nature conservation (the address will be in the Telephone Book) and explain the problem. They will put you in touch with their own bat group. Someone will visit you to examine the roost and then arrange to call back at a time when the animals can be moved with the least disturbance to them.

Bats are perfectly harmless and very useful. They will not tangle themselves in your hair nor bite you. They do not gnaw paintwork, woodwork, electrical

To make a **BAT BOX** for a north facing wall, you will need a pencil, ruler, saw, hammer, nails, & a length of wood 150 × 15 × 2·5 cm.

DO NOT USE WOOD PRESERVATIVES

SIDE · SIDE · FRONT · ROOF · BASE · BACK

All joints nailed

This cut at a slant

Entry slot 2cm wide

Backboard roughened to provide grip

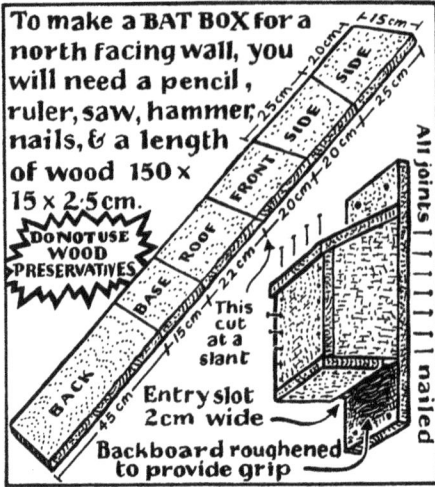

insulation or anything else, and they do not build nests. They do leave droppings, but these soon dry to a powder and cannot harm humans. British bats are insectivores; an individual may take 3,000 midges in a single evening, and they will also eat any wood-boring insects they come across as they rummage around your roof timbers.

You can try to encourage bats by making or buying (ask your trust for nature conservation about it) a 'bat box'. This is a small nesting box made robustly from stout wood but quite simple, with a narrow opening at the bottom. Fix it to a wall, tree, or other convenient object just high enough to deter ground predators, and inspect it once a week or so. It may attract mice or even tits, but if you are lucky a few bats will find it attractive and roost in it. Loss of roosting sites is the main threat facing bats, so bat boxes serve a useful purpose.

If you find a bat that seems to need help, because it is hurt perhaps, you can improvize a different kind of bat box. Find a wooden box that you can seal to prevent the bat escaping and getting into more trouble. Inside the box hang some heavy dark-coloured cloth with plenty of folds in which the bat can hide itself. It will hitch itself to the cloth and dangle darkly until it is better, though you should be warned that the chance of such a bat surviving is not high. If it has injured a wing so it cannot fly, you can take it out in the evening, hitch it to a scarf and wear it. The bat will enjoy being near the warmth of your body, for bats are very social animals.

You can feed an adult bat on scrambled egg, milk, finely minced meat or canned cat food.

Foxes People admire foxes for their beauty, hate them for their cruelty, and are suspicious of them for their cunning. In fact they have surrounded the fox with so many myths that the truth can come as something of a shock. They are related to dogs, but not closely. Cubs can be tamed, and then grow into delightful, playful pets that will romp happily with puppies. They are carnivores and will kill lambs and poultry although they feed mainly on rabbits and rodents, but they also eat large amounts of plant material, especially fruit and berries, and have learned to scavenge from dustbins. Their 'cunning' is what a scientist would call 'behavioural flexibility' an ability to adapt to change by moving into urban areas as their rural habitats deteriorate.

It is this flexibility that may bring a fox into your garden. It will not harm you, but it may give the cat a fright. Their 'killing frenzies' have earned them a bad reputation. What seems to happen is that a fox breaks into a hen-house intending to take one bird as food. This causes all the birds to panic. Normally all but one of them would escape, but they are trapped inside the house. Their flapping and noise trigger the killing behaviour in the fox, which goes on killing until all the birds are dead, though it takes only the one bird it came for in the first place and may lose even that if it is disturbed or frightened.

Hedgehogs You may hear hedgehogs before you see them. They are given to holding noisy family midnight parties with much loud grunting. They feed on insects, slugs and snails, and appreciate a saucer of milk (with which they make an appalling mess), but you will not bribe them to stay in your garden once they make up their minds to leave. They are restless beasts, and just up stakes and go.

Some people are bothered about the fleas that hedgehogs carry. All wild birds and mammals support vast flea populations, but the fleas are specialists and cannot feed on human blood. Hedgehogs have no more fleas than any other animals, but because their hairs are modified into rigid spines there are spaces between them, and the fleas are easier to see.

Hedgehogs hibernate, but may wake up several times during the winter and move to a new nest. Never disturb a hibernating animal; you might well kill it if you wake it. If you do have to pick up a hedgehog let it roll itself into a ball, then insert a long pencil between its belly and legs. It will grip the pencil tightly enough to be carried.

Mice There are six species of mouse resident in Britain: the yellow-necked mouse, harvest mouse, common dormouse, edible dormouse, wood mouse, and house mouse.

The yellow-necked mouse is confined to parts of Wales and southern England and has a distinctive yellow collar. It wanders into gardens provided they are not too far from fields, woods and hedgerows, so you might see one.

Harvest mice, which are tiny, live among tall dense vegetation such as reedbeds, but they have been known to enter farm buildings and are very easy to keep in captivity.

The common dormouse spends most of its time above ground. If you see one you will know it by its bushy tail.

The edible dormouse is confined to parts of Hertfordshire, where it enters houses and can cause damage to apples stored in attics. It looks like a small squirrel with a very bushy tail.

The mice you are most likely to see are the wood mouse and the house mouse. The native wood or field mouse is light brown, has large ears, and looks very neat and clean. It may enter outbuildings or even the house in search of food, but reluctantly, and although it can be kept in captivity it cannot really be tamed.

The house mouse, introduced to Britain as much as two thousand years ago, is grey, eats anything a human will eat, lives in houses and outbuildings, and is the ancestor of the pet white mouse. Those you find in towns are much better adjusted to the presence of humans, and enter houses much more readily than those living in rural areas. These are the 'town mouse' and 'country mouse' of children's stories. The house mouse is regarded as a pest, but provided food is kept in proper containers it is harmless.

Never pick up any mouse by the tail. The mouse can turn around, climb its own tail and bite your finger; more seriously, the skin can slide off the tail, which is then likely to become infected and the mouse can die rather nastily.

Moles Apart from their molehills, which you may find unsightly if they occur on your lawn, moles are harmless beasts that enjoy a solitary myopic life feeding on worms and other small animals that drop into their galleries and are collected during routine patrols.

Moles dislike one another, so you are unlikely to have more than one in the garden, and they prefer undisturbed fertile soil with plenty of little animals in it.

You may be lucky enough to see your mole if you go out at night with a torch. They often surface to take the evening air, sniffing carefully to check that no predators are around before they emerge. The mole will not be troubled by a light, but it has possibly the keenest sense of smell in the entire animal kingdom, so unless you are careful it will smell you long before you see it. Wear thick gloves if you have to pick up a mole — they bite!

Rabbits Rabbits were introduced into Britain from the Mediterranean region, probably in the twelfth century and perhaps by returning Crusaders, but it was not until about a hundred years ago that their population 'exploded' as new farming methods divided previously open and exposed country into arable fields bordered with hedges. This provided the rabbits with an almost inexhaustible supply of food, cover and sites for burrowing, a habit unique to rabbits (neither hares nor the North American cottontails excavate burrows in this way). These days the rabbit population fluctuates locally as

the animal comes to terms with the myxomatosis virus — the virus mutates to a virulent form, rabbits die, the survivors recover, the population increases, and then the virus mutates again, so up and down goes the population.

Rabbits may visit your garden, but they are wary, shy creatures of the twilight and you may not see them, though you may find their footprints in snow or soft earth. They feed on grass and other vegetation. Domesticated rabbits are descended from the ordinary wild rabbit, but if they escape they are unlikely to survive.

Rats Everybody hates rats, and if the neighbours find out that you have rats around the place they will hate you too. People from the council will come and kill them. Rats do indeed damage stored food, but the harm they do is greatly exaggerated.

There are two species of rat: the ship or black rat (which is not necessarily black), and the larger common or brown rat. Neither is really a native. The ship rat has lived in Britain since pre-Roman times, the common rat since the eighteenth century, and the common rat has largely displaced its predecessor, which survives only in some dockyards and is rare even there.

Rats are intelligent gregarious animals with a highly developed social structure and a genius for making a living under trying circumstances. They will eat anything a human will eat, but are less fussy, and from their wild origins they have become adept at living in and around the sewers beneath human dwellings. If you see the odd rat take no notice. If you see several and they appear to be residents, you can either study their behaviour or scream for help.

Shrews If you have an area of dense vegetation and long grass, you may be lucky enough to come across a mouse-

like animal (though smaller than a mouse) with dark fur and a very long, very mobile nose. You may even see it sit up to sniff the air. It will be a shrew. If it is brown and really tiny it may be a pygmy shrew, one of the smallest of all mammals.

Shrews eat insects and small invertebrates, and sometimes tunnel through the grass or just below the soil surface. Being so small, they must spend most of their time feeding just to provide enough energy to maintain their body temperature, so do not interfere with them because they have no time to spare arguing with you. They produce several litters a year. Like most mammals, the mother shrew moves her infants to a new nest as soon as they are mobile, but her method is unique. She lines up the youngsters, each holding on to the tail of the one in front, with the front one holding her tail. Off they go in a caravan, and if one lets go she retrieves it and pushes it back into place none too gently.

Squirrels The common grey squirrel was introduced to Britain from North America less than a century ago. It is now much more numerous than the native red squirrel, but the rise of one species is not the cause of decline of the other — red squirrel populations have always fluctuated widely.

Grey squirrels are often patchily brown or chestnut in colour, especially in summer. They live in broadleaved woodland, the red squirrel preferring coniferous woodland, and both feed on seeds and nuts, though they will also eat shoots, leaves and fungi. Grey squirrels spend more time on the ground than red squirrels, and in town parks they can become very tame, even allowing people to feed them by hand. Many people regard the grey squirrel as a pest, but it will do no harm in your garden.

Voles If you live in the country, the mouse-like animal you see in the garden may be a vole. It looks rougher and more unkempt than a wood mouse, and has a shorter tail. If it is slightly chestnut in colour and the tail is about half as long as the head and body together, it is a bank vole, and will have come from nearby woodland or a dense hedge. If it is greyer or yellowish, has a very blunt snout and a short tail, it is a field vole from some rough grassland not far away.

If a slow-moving river flows through or beside your garden, or you have a ditch that always has water in it or a large old pond, you may see a rat-like animal swimming in the water. It may be a rat (rats do swim if they have to), but it may also be a water vole, a mainly herbivorous animal about twenty centimetres long. If you get a clear view of it you can tell which is which — if the tail looks long and naked, the ears large and the muzzle long, it is a rat; if the tail looks short and hairy, the ears small and the face short, it is a water vole. 'Ratty' in *The Wind in the Willows* is really a water vole.

Conclusion

So Where From Here?

Ours is by far the most highly urbanized civilization the world has ever known. If you look back across the millennia of history you will see farmers — peasants, most of them — wresting a miserable living from an impoverished acre or two, people fishing, foresters and rural labourers, all of them sweating long hours to supply themselves and a tiny urban population with the necessities of life. Our ancestors were country folk, even though some of them lived in cities. Until quite recently even the cities were quite rural, and farm livestock was a common sight. I can remember real bulls being marketed in Birmingham's Bull Ring.

Yet I was not a country person. The city was my home, the place in which I knew how to live. My environment included factories, railways, buses, busy streets and vast numbers of people. The most recent statistics show that 92.5 per cent of British people live in urban areas. They ensure that our entire culture is an urban one, so that even the 7.5 per cent of us who live in the country cannot avoid being influenced by urban values. Even the country dwellers are half-way to being urbanized, and depend on cities for their goods and services just as much as the city dwellers.

This is partly why it is so easy to hear or read about 'environmental problems' and think of them as somehow remote. Conservation of wildlife, protection of the landscape, acid rain — all seem essentially rural concerns, seen through urban eyes. They may be problems, but they are not our problems.

I hope this book has persuaded you that the environment affected by these problems is the environment that we all share. We affect it simply by living in it, and we in turn are affected by it. The dark forests which alter landscapes and habitats, and which are suffering from acid rain, were planted to supply city dwellers with timber and timber products and so reduce import bills. The farmed landscape which contributes to the pollution of the environment by pesticides and nitrates is dedicated to producing food for city dwellers to eat. Though we may live in cities, we all bear some responsibility for the rural environment.

Our cities and conurbations are enormous. A glance at the map will show you the size of Greater London, the West Midlands, the Lancashire and Yorkshire cities, the Central Lowlands of Scotland. Yet the same glance will also show you that large though they are, they occupy only a small part of the total land area. Urbanization means that we live crowded together in a small space — that, too, is an environment, and one in which we affect each other directly. Environmental sensitivity in towns is a matter of common courtesy to our neighbours, of elementary good citizenship.

This has been a book of hints, suggestions and information intended to help you distinguish those problems that are real from those that are not, and those about which you can (and perhaps should) do something, and those which are outside your influence. You cannot solve problems of poverty and hunger in Africa (although you might put pressure on politicians who could at least help solve them); you can reduce air pollution by burning less fuel. You cannot save the world's rain forests single-handed, but you can provide sanctuary for local wildlife in your own garden. If this book has persuaded you that you have more control over your own way of life and your environment than you may have supposed, you will probably wonder where next to turn, how to pursue the matter further.

I have not compiled an extended reading list of titles for you to chase through bookshops and libraries, though I have suggested a few very practical titles. There are many books about the environment, but if you are to come to grips with the issues then you will need to dig deep, below the generalities, and this means tapping sources of specialist scientific information. If you start to do that you will soon discover that there is no end to it. A suitable list of titles would rapidly get out of control and grow longer and longer.

There is another way to start. Ask questions, meet people, get in touch with relevant organizations. If you come across a particular environmental problem, especially one that affects you directly, try going straight to someone whose job it is to deal with it. Your local authority will be able to help with a wide range of matters, and it is always worth trying them first. If you are worried about radiation, try writing to the National Radiological Protection Board; if you want a particular activity to stop, contact one of the environmental organizations. If you want to do more to help, encourage or understand wildlife, join your local trust for nature conservation — there is one in each county. If you cannot find the local address, write to the Royal Society for Nature Conservation who will give you the address of your county trust.

Most of the large national organizations have local branches, and you should be able to find their addresses easily enough in the telephone book or at your local library. If this fails, write to the head office whose address is given in the address list and ask for the address of the local branch (always enclose a stamped addressed envelope).

There are a few glaring omissions from the list of addresses, but they are unavoidable. The quality of water in the ground, in rivers and lakes, and in your taps, is the responsibility of the relevant water authority, but this will change when the water industry is privatized. The present plan is to establish a national authority to supervise water quality, but no one yet knows how this will operate in practice. It is possible that the national authority will contract its tasks to the private companies which will control water supply.

Plans to privatize the electricity industries may result in privately-owned power stations being regulated just like ordinary factories. This would presumably include the nuclear power stations, though no one knows how this would affect that part of the nuclear industry supplying fuel and disposing of wastes.

Suggested Reading

Allaby, Michael *Ecology Facts* Hamlyn, 1986

Allaby, Michael *Making and Managing a Smallholding* David and Charles, 1986 (2nd edition)

Carr, Donald E. *Energy and the Earth Machine* Abacus, 1978

Cross, Michael *Grow Your Own Energy* Blackwell, 1984

Gray, Juliet *Food Intolerance: Fact and Fiction* Grafton, 1986

McCartney, Kevin and Brian Ford *Practical Solar Heating* Prism, 1978

McGuigan, Dermot *Small-Scale Water Power* Prism, 1978

McGuigan, Dermot *Small-Scale Wind Power* Prism, 1978

McLaughlin, T.P. *A House for the Future* TV Times, 1976

Martin, Alan and Samuel A. Harbison *An Introduction to Radiation Protection* Chapman and Hall, 1979 (2nd edition)

National Radiological Protection Board *Living With Radiation* 1986 (3rd edition)

Owen, Jennifer *Garden Life* Chatto and Windus, 1983

Rowland, Anthony J. and Paul Cooper *Environment and Health* Edward Arnold, 1983

Yudkin, John *Penguin Encyclopedia of Nutrition* Penguin, 1985

If you are looking for books about Britain's wildlife, Collins' new *Field Guides* are excellent and are very ecologically based, dealing with habitat and lifestyle and not merely identification.

For Product Safety Concerns and Information please contact our EU
representative GPSR@taylorandfrancis.com
Taylor & Francis Verlag GmbH, Kaufingerstraße 24, 80331 München, Germany